WINN LIBRARY
Gordon College
Wenham, Mass. 01984

W9-AFN-080

QH
325
.D34

DISCARDED
JENKS LRC
GORDON COLLEGE

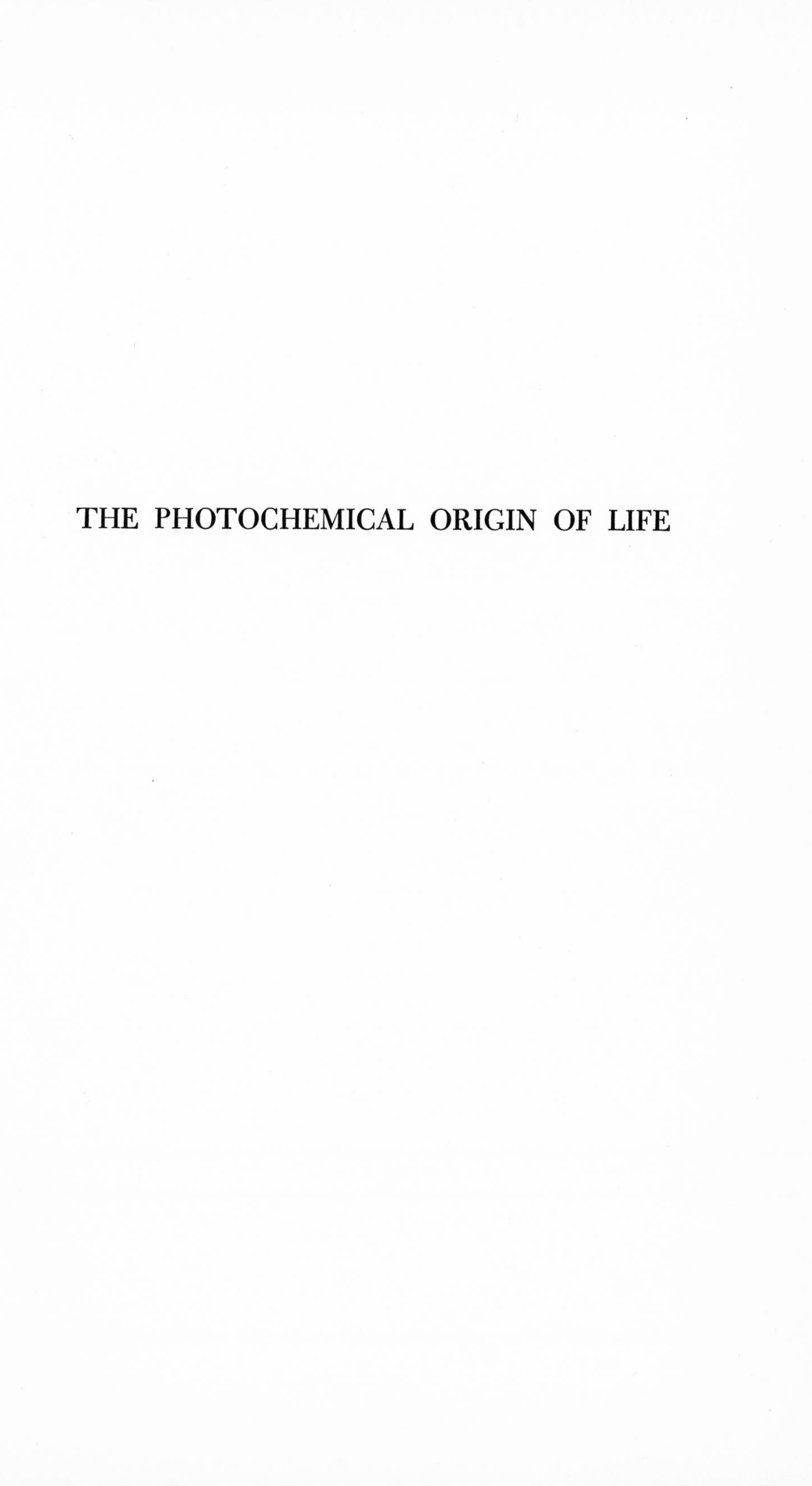

THE PHOTOCHEMICAL ORIGIN OF LIFE

"Retrace the history of the Universe from the birth of atoms. Show how, under the influence of gravitation, they were concentrated into Nebulae, Stars and the whole multitude of celestial bodies. How, on at least one planet, they combined to form organic compounds. How these organic compounds became living matter. How these living creatures, as they evolved, gave birth to a superior race. How, after many centuries, a handful of learned men who had come from this superior race, glancing around themselves at this world which, blindly, had given birth to them, judged it and came to know it for what it is. . ."

J. A. BALFOUR BRITISH ASSOCIATION FOR THE ADVANCEMENT OF SCIENCE
CAMBRIDGE (1905)

THE PHOTOCHEMICAL ORIGIN OF LIFE

by A. Dauvillier

OBSERVATOIRE DU PIC-DU-MIDI
LABORATOIRE DE PHYSIQUE COSMIQUE
BAGNÈRES-DE-BIGORRE, FRANCE

Translated from the French by SCRIPTA TECHNICA, INC.

1965

ACADEMIC PRESS New York and London

QH
325
.D34

COPYRIGHT © 1965, BY ACADEMIC PRESS INC.
ALL RIGHTS RESERVED.
NO PART OF THIS BOOK MAY BE REPRODUCED IN ANY FORM,
BY PHOTOSTAT, MICROFILM, OR ANY OTHER MEANS, WITHOUT
WRITTEN PERMISSION FROM THE PUBLISHERS.

ACADEMIC PRESS INC.
111 Fifth Avenue, New York, New York 10003

United Kingdom Edition published by
ACADEMIC PRESS INC. (LONDON) LTD.
Berkeley Square House, London W.1

LIBRARY OF CONGRESS CATALOG CARD NUMBER: 65-15771

PRINTED IN THE UNITED STATES OF AMERICA.

First published in the French language under the title
L'ORIGINE PHOTOCHIMIQUE DE LA VIE and copyrighted in
1958 by Masson et Cie, Éditeurs—Paris, France

PREFACE

The problem of the origin of life is as old as humanity; it is one of the most important questions of natural philosophy. For thousands of years, philosophers and religious leaders offered solutions within the limits of the knowledge of their times; contemporary scientists, however, are more exacting. Our knowledge has advanced so far that plausible solutions could have been formulated since the beginning of this century.

The problem is not a biological one, since living beings did not exist in the beginning. It is truly an interdisciplinary problem of cosmic physics involving astronomy, geophysics, and geochemistry. We are going to show that the cosmic paleovulcanism, which violently disturbed the surface of our planet from its beginning, played an important role by bringing forth the pyrogenetic synthesis of numerous heterocyclic compounds. These heterocyclic compounds still exist on the surface of Mars and the moon, as shown by samples of cosmic rocks from certain meteorites. We will also see that after the condensation of the oceans, when our atmosphere resembled those of Mars and Venus, photochemical reactions took place on the seawater's surface that were capable of creating the optical rotation characteristic of living matter. In the same way that the sunlight maintains life on the surface of the earth, it is also responsible—although in another spectral range—for the appearance of the primitive organic matter from which living material was derived.

The study of the life on earth is the most effective method at our disposal for an understanding of cosmic life. Although the existence of cosmic life has not been and most likely will never be recognized in our solar system, it is probably extremely well known in the galaxy. However, since the interstellar distances are so great, it is feared that we will never obtain confirmation of this cosmic life.

This theory of the origin of life was worked out by Dr. E. Desguin and the author in 1938 and is the subject of several books published in French. One of the books was translated into Italian and published by Feltrinelli in 1962.

The author is happy to present this first English translation to the

American public. Since the time of W. M. Stanley's famous works, we have been witnessing in America a prodigiously active scientific development instigated by a magnificent youthful enthusiasm which we admire and envy.

This new edition allowed us the opportunity of adding a number of new facts that were not given in the original text published in 1958 by Masson of Paris. The author fully appreciates the quality of the translation which faithfully interprets his thoughts. He also thanks Academic Press for the careful presentation of his work.

A. DAUVILLIER

April, 1965

CONTENTS

PREFACE v

INTRODUCTION

References 13
Summary of Definitions, Nomenclature, and Numbers of Physical Significance 14

I. CHARACTERISTICS OF LIVING MATTER

Chemical Composition 16
The Chemistry of Carbon 20
Molecular Dissymmetry 21
Macromolecules 22
Living Creatures 34
Action of Physical Agents 39
References 43

II. THEORIES OF THE ORIGIN OF LIFE

Spontaneous Generation 47
Cosmic Panspermia 49
Creative Chance 53
The Synthesis of Organic Matter 57
References 59

III. COSMOGONY AND GEOGENY: The Origin of the Earth

The Evolution of the Universe 61
The Origin of the Solar System 63
The Astronomical Problem 65
The Mechanical Problem 66
The Physicochemical Problem 67
References 69

IV. THE ORIGIN OF THE CONTINENTS, OCEANS, AND ATMOSPHERES

The Cosmic Chemistry of the Lithosphere 70
The Pyrogenic Syntheses of Organic Compounds 73
The Formation of Water Vapor and of Carbon Dioxide. The Origin of the Oceans 75
The Primitive Marine Environment; The Volcanic Cycle of the Salt 85
The Origin of the Atmosphere 88
References 91

V. THE PHOTOCHEMICAL SYNTHESIS OF ORGANIC MATTER

The Origin of Optical Activity 107
References 115

VI. THE ORGANIZATION OF LIVING MATTER. THE EVOLUTION OF LIVING CREATURES

The Appearance of Chlorophyll 122
Unicellular Organisms 129
The Evolution of Living Creatures 132
References 142

VII. THE ENERGETICS OF THE BIOSPHERE

Text 144
References 151

VIII. THE GEOCHEMICAL ROLE OF THE BIOSPHERE

The Carbon Cycle 156
Nitrogen 160
Phosphorus 162
Sulfur 163
Manganese 164
Iron 165
References 166

IX. LIFE IN THE UNIVERSE

Text 168
References 177

Bibliography 178

Subject Index 181

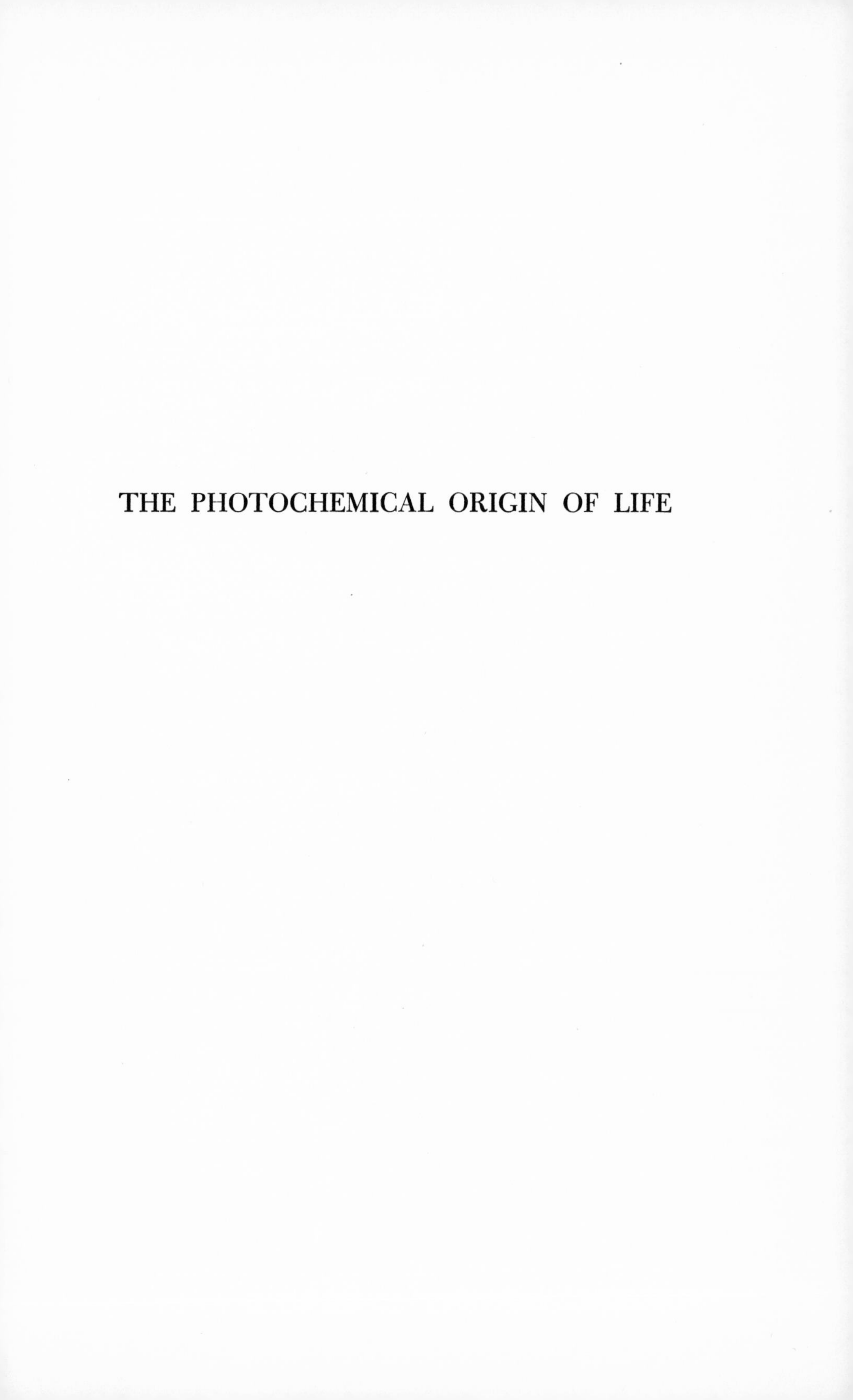

THE PHOTOCHEMICAL ORIGIN OF LIFE

INTRODUCTION

The problem of the origin of life has only recently been tackled in a scientific manner since the work of Pasteur. It was he who, in discovering the molecular dissymmetry of living matter, looked most deeply into the nature of life. The ancients assumed that small living creatures could be born spontaneously, without parents, from the four elements of Empedocles. It required secular controversies, and the work of Schwann, and then of Pasteur, to refute the doctrine of spontaneous generation and prove true Harvey's adage (1619): "omne vivum ex vivo."

But life is a phenomenon so marvelous, so apart from all others, that man, discouraged by its complexity, could only attribute it to the supernatural. Man's fearful imagination peopled the earth and the heavens with hidden powers who controlled the birth, evolution and death of all life.

In 1790, I. Kant still held a supernatural and finalist conception of life. Yet, more than a century before, Descartes had already—in a slightly simplistic way—regarded life as purely mechanical. Huygens, in his "Kosmotheoros," at the dawn of the eighteenth century, was the first to consider life as a universal, cosmic phenomenon.

This conception was held by a number of learned men in the last century. For Liebig, Lord Kelvin, and Pasteur, life was a property of matter, as universal and eternal as the world itself. For H. von Helmholtz, Kelvin, and Arrhenius, life did no more than change position in space on passing from one celestial body to another. This doctrine of *cosmic panspermia* is not without grandeur. It postulates that life is everywhere analogous to terrestrial life, since the latter derives from the common source.

With this conception, the problem at once assumes a cosmic, astronomical character. If life is as eternal as the universe itself, the problem of its beginning no longer arises; it is a false problem. But a cosmic mechanism must be found that is capable of seeding the heavenly bodies. We shall see that no adequate process has been or could have been proposed.

Pasteur's concept of a life inherent in matter is still very widespread among a great many philosophers, naturalists, and biologists, for example, A. Comte, Bergson, E. Leroy, and Teilhard de Chardin. Many philosophers are fond of talking of the "life" of metals, of atoms, even of particles, thereby lending them hidden properties. An erudite historian declared that he was persuaded that life began in the ocean depths. Now, clearly life did not begin in the arid deserts or on the polar ice, or on the high mountains, or in the dark frozen abysses, since all these regions had to be subsequently conquered by a slow adaptation. The literary man too often believes without sufficient justification. An unbridgeable chasm will always exist between the thinking of the man of science and that of the man who was not initiated into the practice of accurate observation and experimental method.

Cosmic panspermia, in its least unlikely form, appeals, with H. E. Richter (1865), to the carriage of germs by meteorites. Now, we know today, from their isotope composition, that the cosmic rocks belong to the solar system and have their origin in the asteroid ring, i.e., the debris of the dead planet of Olbers which broke up at the "lunar" stage. Not only do the cosmic rocks contain no mineral of sedimentary or biogenic origin, but their carbides, graphites, and diamonds testify to an igneous origin. We have no reason to believe in the existence of such rocks in interstellar space, since their isotope composition would be different and we have never obtained any for observation. The bacteriological research of Pasteur, carried out on carbonaceous meteorites, led, and for a very good reason, to negative results.

Certainly life is scattered in abundance throughout the galaxy and is as eternal as the universe itself. But the cosmic environment is so full of hazards, and the interstellar distances are such, that planetary germs—assuming, against all possibility, that they could escape from their planet—would actually have no chance of reaching us. If they did so, they would still have to find, by some extraordinary fluke, a planet with extremely improbable receptive conditions that shall be specified. As a matter of fact, the doctrine of cosmic panspermia can only be conceived if one accepts the idea of the carriage, spontaneous or otherwise, of live germs by foreign astronauts. This, to all intents and purposes, is fantasy. Cosmic panspermia is a facile hypothesis, a subterfuge which seeks to avoid the fundamental problem of the origin of life.

This problem consequently assumes, on first sight, a cosmic aspect, and, in order to deal with it, it is necessary to begin with a plausible cosmogonic conception. We can no longer conceive, with Laplace, of an eternal, immutable world, an unchanging solar system, an inexhaustible and indefinite source of heat and light the sun, and an eternal terrestrial life without beginning and end. Today, we see the stars and galaxies evolving rapidly on the cosmic time scale, where the unit is a billion years. Whatever hypotheses we adopt for the evolution of the universe and the formation of the planetary systems, observation of the planets nearest to earth, Venus and Mars—apparently sterile, as we shall see—shows us the conditions that prevailed on a primitive, early cooled earth. Thermochemistry, the application of the kinetic theory of gases to the planetary atmospheres, and the study of the evolution of magma during its cooling, enable us to gauge the geogenic and geochemical evolution that led to the formation of the continents, oceans, and atmospheres. We find that a primitive earth, just like Mars and Venus, was unfit to receive life. The atmosphere of nitrogen and carbon dioxide of volcanic origin was transparent to the abiotic solar ultraviolet. Any living creature transported to these planets could only perish because of lack of nourishment, oxygen, and protection against ultraviolet light.

Life, therefore, certainly began on earth by spontaneous generation some thousands of millions of years ago, just as it must appear on any planet whose mass is sufficient to retain a hydrosphere and an atmosphere, and is situated at a proper distance from its mother star and is rotated with sufficient rapidity for the water on it to remain liquid. But life probably appeared in a different, very simple, primitive form, that is, on the scale of the laboratory of nature and of geological time. The question here is a very general one of cosmic physics and chemistry and for some decades we have had all the data required to deal with it fruitfully.

It seems that the first scientific hypothesis formulated on the subject of the origin of life should be attributed to J. B. Lamarck (Philosophie Zoologique, 1809). But in 1863, Darwin still felt it was "unreasonable" to think about the origin of life. After Winogradsky's discovery of the autotrophic bacteria in 1890, a number of authors, H. F. Osborn, Ed. Perrier and Constantin, suggested that life began, on a primitive sterile earth with these bacteria. Now these bacteria are much more complex than the heterotrophic bacteria since they are endowed with more extensive powers of synthesis. It can be shown that their energy balance would be weakened. They represent an advanced stage in evolution and are considered today—leaving the ferrobacteria aside—as parasites utilizing the waste products of life, ammonia, hydrogen sulfide, and methane. The same objection prohibits ascribing the commencement of life to the almost miraculous appearance of a grain of chlorophyll, since chlorophyll is extremely complex and requires, as B. Moore noted, a living substrate produced by a long previous evolution.

In our time B. Moore, R. Esnault-Pelterie, and Prenant have shifted the doctrine of spontaneous generation back into the submicroscopic region, to the virus scale. So this doctrine is retreating step by step as observation improves and this observation has now arrived at the limit of the living world, beyond which the organized being gives place to the macromolecule. Then, the creative role of chance may be considered. This is the ancient thesis of the progressive exhaustion of fortuitous combinations. "Length of time may bring anything to pass," Herodotus pointed out. The advent of probability calculation, Pasteur's discovery of the role of asymmetric structures, and the study of the macromolecules by Staudinger, Pauling, Mark, and Wyckoff led C. E. Guye, in 1922, to examine the role of chance. He showed that thermal molecular agitation—the activity of Brownian movement—was totally incapable, on the available time scale, of giving rise, in an inorganic world, to organic macromolecules of asymmetric structure.

Other learned men have, more properly, sought for mechanisms that would cause abundant, sterile, organic substances that would later become living matter, to appear on an inorganic earth. However, although we observe the constant formation of thousands of minerals, we never see organic compounds as simple as tartaric acid form in nature itself. All the natural organic substances known to us—coal, lignites, petroleum, and resins—are products or waste materials of life. Many rare iron and calcium oxalates and a mellate of aluminum are encountered, but they are thought to be due to the oxidation of lignites. The rare carbonaceous substances brought by some meteorites, the so-called "carbonaceous meteorites," probably resulted from the action of superheated steam on hydrocarbons. The reason for this is that organic substances, being endothermic, require, for their synthesis, external terrestrial or cosmic energy, such as lightning, the high temperature of volcanic chambers, radioactivity, or ultraviolet light from the sun. In 1875, Pflüger suggested that cyano-compounds were synthesized pyrogenically during the globe's period of incandescence, and A. Gautier saw the origin of life in such compounds produced by vulcanism. Nowadays, the traces of formic acid and formaldehyde produced by volcanoes are unstable in the presence of atmospheric ozone and never accumulate. There is no doubt, however, that aromatic nuclei first appeared during a primitive era of pyrogenic syntheses. The hydrocarbons, such as acetylene, were able to react with sulfur, ammonia, and hydrogen cyanide, to give thiophene, pyrrole, and pyridine, which played an important role in the process of time.

It is altogether remarkable that, in 1873, E. du Bois-Reymond was already attributing these syntheses to the ultraviolet solar rays. This idea was taken up again by B. Moore, and then by P. Becquerel, who placed these syntheses in the upper atmosphere. For the moment, the problem seemed insoluble, since, to effect the synthesis of ternary and quaternary organic compounds at the expense of carbon dioxide, steam, and ammonia, it was necessary to resort to high-quantum ultraviolet radiations, to which our atmosphere was opaque. Moreover, such syntheses, taking place in the upper atmosphere in a gaseous medium, would lead to the immediate photolysis, by the near ultraviolet, of the compounds formed.

Ultraviolet light, through its high quantum, is capable of effecting, at ordinary temperature, reactions, such as the dissociation of water and carbon dioxide, which require a high temperature. It is eminently suitable for carrying out the synthesis of complex organic substances which would be destroyed by a slight rise in temperature. As the works of Pasteur and P. Curie have shown, it alone is capable of bringing about the synthesis of the asymmetric molecules characteristic of living matter. Life is the result of molecular dissymmetry, just as our world is the result of the electrical dissymmetry of the elementary particles.

Thus we come to the idea that the problem of the appearance of organic matter is purely an interscience problem of cosmic physics. We must explain, at one and the same time, the transparency of the primitive atmosphere to the far ultraviolet, the photosynthesis of ternary and quaternary compounds and asymmetric molecules, and the appearance of an atmosphere of free oxygen that would permit life to survive while protecting it against the abiotic action of these radiations.

All the elements of this problem were known in 1910, but it was only in 1938 that we knew how to connect them. In 1865, Köne had shown that the primitive atmosphere of the earth was unable to contain either oxygen or, consequently, ozone. Oxygen is the product of the chlorophyll function, and, if life ceased, it would soon disappear in oxidizing nitrogen under the influence of lightning, in burning combustibles, volcanic effluvia, such as hydrogen and sulfur, and in oxidizing ferruginous lava and a number of minerals, such as the metallic sulfides. That is why the atmosphere of any planet does not reveal the presence of traces of oxygen, the characteristic substance of life. The oceans and rocks of the lithosphere are produced by the oxidation of hydrogen, silicon, and the alkali earth elements. The terrestrial planets are the *cosmic ashes* of the combustion of "Russell's solar mixture." The primitive oxygen-free atmosphere was of the same character as of Venus and Mars, that is to say, rich in carbon dioxide and nitrogen from paleovolcanic activity. It was transparent to the high-quantum far ultraviolet. Although hydrocarbons are absorbed in the ultraviolet which would prevent all photosynthesis, carbon dioxide is transparent up to 1800 Å. Vulcanism produces mainly carbon dioxide but hardly any hydrocarbons. The methane soffioni are of biogenic origin. Although hydrides are characteristic of the giant planets because of their mass, oxides are characteristic of the terrestrial planets. It is possible to say, in the vivid words of J. Pompecky (1925): "Without radioactivity, no vulcanism; without vulcanism, no carbon dioxide; without carbon dioxide, no life." The entire carbon of the living world is, in fact, derived from dissolved or atmospheric carbon dioxide.

It is oxygen in all its forms that is the main atmospheric absorbent. Atomic oxygen is one of the absorbents of the ionosphere. Molecular oxygen, O_2, absorbs, in a thickness of a few centimeters, all radiations shorter than 2000 Å (Schumann-Runge bands). It is ozone, O_3, which, in a thickness of a few millimeters of normal gas, limits the solar spectrum to 3000 Å, as Hartley showed in 1880.

The photosynthesis of organic compounds was first achieved in the laboratory by H. Slosse in 1898 at the Solvay Institute in Brussels using the corona tube of M. Berthelot. He obtained sugars. The corona tube is a very powerful synthesizing agent since the radiation is produced in the gas itself in high yield without any parasitic absorption. It is not pure photochemistry because the gas is ionized. Moreover, a large part of the synthesis products are decomposed by the discharge and undergo photolysis due to the near ultraviolet. But, in 1910, in a very important piece of work carried out with the aid of the quartz mercury arc, D. Berthelot and H. Gaudechon synthesized ternary and quaternary compounds from carbon dioxide, steam, and ammonia.

Therefore, these reactions could have taken place at the surface of the primitive waters. They result in the production of glucides: aldehydes, alcohols, sugars, and complex quaternary compounds derived from formamide. Just as carbonic acid and water combined to give carbohydrates, so nitrogen and water combined, under the influence of lightning and the extreme ultraviolet, to produce ammonium nitrite. This is a reaction which, in our view, still occurs in the atmosphere of Venus, and is the cause of the yellowish white clouds that cover the planet. The simultaneous photochemical synthesis of formaldehyde and formamide leads to glycine, the second substance in the amino acid series, the bases of proteins.

In 1860, Pasteur, in discovering the molecular dissymmetry of the organic constituents of living matter, at first thought that optical activity was characteristic of life and that organic synthesis was powerless to reproduce these compounds artificially. In the same way, before Wöhler's synthesis of urea in 1828, it was believed that the elaboration of organic compounds was the exclusive prerogative of life. However, in 1873, E.

Jungfleisch succeeded in synthesizing the resolved racemic compound of tartaric acid, and in 1886, A. Ladenburg succeeded in synthesizing cicutine or conine, the alkaloid from hemlock. But, it was still not true total asymmetric synthesis since the two enantiomorphs were produced in equal quantity, although plants produce only one optical antimer. Total, asymmetric, pure photosyntheses were achieved subsequently by the action of circular polarized light.

In nature itself, the asymmetric molecules characteristic of life are produced by a catalytic action of diastases, which are themselves asymmetric. The active molecules derive from each other, as do crystals and living creatures, and the problem of the origin of the first generative molecule is inseparable from that of the origin of life. Becquerel, Byk, and Vernadsky suggested a number of hypotheses to account for the catalytic action of diastases, but only one, the action of circular polarized sunlight (Le Bel, Van't Hoff), through some fortuitous crystalline optical assembly, has survived. The first asymmetric synthesis of a macromolecule capable of self-reproduction took place then, on the molecular level, alongside some salty lagoon surrounded by quartz and calcite crystals.

Molecular asymmetry, the basis of life, would thus have depended on the previous asymmetry of the crystal lattice, and this condition is certainly in keeping with geochemical evolution. Crystallization and life are two aspects of the organization of matter, of the appearance of order in molecular chaos, but whereas crystallization reveals a static coordination, life reveals a dynamic one (E. Desguin).

This asymmetric synthesis occurred once only. The asymmetric molecule, as soon as it was created, imposed its structure on the organic environment, just as the crystal seed leads directly to the crystallization of the saturated solution in which it is sown. The probability of the appearance of inverse symmetry was the same. Thus, on each planet containing living creatures, there exists only one type of enantiomorph. Never, in the course of time, was any synthetic organic matter produced in a sufficient isolated mass for further asymmetric synthesis to be able to take place. In the same way, no trace of sterile organic matter has a chance of subsisting in real nature without immediately becoming the prey of the living world.

Asymmetric synthesis applies to the glucides as well as to the protids. Photochemistry alone is capable of effecting total asymmetric syntheses, and because of this we find a decisive argument in favor of the photochemical theory of the origin of life.

Any theory of the origin of life should also be capable of describing accurately the condition of the primitive marine environment. We have shown (1) how the primitive oceans had straightaway acquired their present salinity and how marine life had been able to persist for nearly four thousand million years, thanks to the "volcanic cycle of the salt."

The works of Boussingault and Schloesing demonstrated the oceans' regulatory role in maintaining the level of carbon dioxide in the atmosphere. The biosphere never risked being asphyxiated by the young carbon dioxide of volcanic origin. The greater part of the nitrogen of the present-day biosphere was then found in the oceans, in the form of ammonium bicarbonate. Therefore, at the surface of the marine waters, we are witnessing the appearance of a new type of geological formation, namely, vast banks of gelatinous substances in a warm environment of sugar and ammonia, and in the presence of ammonium carbonate and phosphate. It is only at this stage of geochemical evolution that a terrestrial planet could be seeded successfully by a foreign cosmic germ. Sea water, through its salinity and alkalinity, is the medium most favorable for the emergence of life. Like the fertile volcanic terrains derived from the magma, it contains all the oligo-elements, known and unknown, which are necessary for life. Geochemistry has brought to light the close analogy existing between the statistical chemical composition of the hydrosphere and of living matter. The elements necessary for the formation of protein and nucleoprotein macromolecules, elements such as sulfur, phosphorus, magnesium, manganese, and iron, are found directly in the marine environment. This organic matter will be in the presence of dissolved oxygen and will be in a *metastable* state. It will not be able to burn or explode since it is in an aqueous environment, but it will have a *tendency* to recombine with free oxygen, by means of a suitable *organization.*

Now, it is just this recombination of endothermal organic substances with the oxygen, and the release of energy, that is the fundamental characteristic and the mechanism of life, the *cause* and *necessity* of which we thus understand. Living creatures can be compared to explosives, except they have the power to control and moderate the oxidation and release of energy that animates them (E. Desguin).

Thus, by a curious paradox, it is to the abiotic far ultraviolet that we turn for life to appear. According to the old conception, life originated in the marine environment from Aristotle's four elements: water, air (carbon dioxide), earth, (some mineral elements), and "fire" (solar radiation).

In what epoch of geochemical history are these photosyntheses placed? From the works of M. Milankovitch, Sir H. Jeffreys and others, it appears that the earth's surface cooled extremely rapidly, in a period of the order of ten thousand years only, since the present-day climatic pattern was established as soon as the surface magma solidified. Less than one million years after the earth's birth, and thus more than four thousand million years ago, the cooled oceans were already ready for photosynthesis. As the recognized geological epochs hardly total more than five hundred million years, there is a considerable period of evolutionary time available before we come to the Protista.

The problem of the origin of life is therefore divided into two very distinct problems. The first, cosmic and geochemical, is concerned with the evolution of inorganic matter into organic matter. The second, much more complex, concerns the elaboration of the structures in this environment and forms the bridge between biochemistry and cytology.

Experience has shown that the chemical complexity of this type of environment will increase with time. The first stage of spontaneous organization will be the appearance of the molecules of adenosine-phosphoric acids and of the many tens of enzymes of *anaerobic fermentation.* This fermentation of the organic matter will make it, according to our energy definition, a *living environment,* although it still will be without organisms, just as the cytoplasm of the actual cell is a living environment.

The adenosine-phosphoric acids will lead to the formation of nucleoprotein macromolecules capable of reproducing themselves, that is to say, of naked "genes." The highly complex molecular organizations which constitute viruses are known to possess this "biological" property. The viruses, which can sometimes be crystallized, are not living since they do not respire and do not contain water, salts, sugars, fats, or any diffusible substance. They are, however, endowed with genetic continuity, and they only reproduce in a host which is itself living. This structural perenniality, which is conceivable in the fairly simple molecule or the crystal lattice, is undoubtedly surprising in a macromolecule with a mass of several millions, but it is an observed fact.

These particles, endowed with genetic continuity, will therefore be able to reproduce themselves at the expense of the proteins of the environment, since they are in a *living environment,* although one without organisms, just as a virus reproduces today in a living host.

A second stage of organization will correspond to the appearance of those "plasma membranes," so thoroughly studied by H. Devaux since 1923. The amino acids form polarized chains with an alkaline and an acid end. They may be attached end to end, like magnets, with the elimination of a molecule of water, forming linear chains, or they may be juxtaposed, forming a lattice whose two faces have different properties. These membranes, constructed of oriented heteropolar molecules, form semiwettable catalytic surfaces. For these surfaces to be coiled into closed sacks, two differing fields of action exist on the surrounding environment, one toward the interior, the other toward the exterior. These closed sacks, of the order of 1 micron in size, formed by a membrane with different catalytic power on the two faces, will have the capacity to secrete substances such as fat, starch, and glycogen through their inner face, and soluble enzymes, such as oxidases, through their outer face. How do they come to contain macromolecules of deoxyribonucleic acid in the role of central body? They will correspond to the outlines of heterotrophic bacteria and mitochondria. The bacteria are the smallest living creatures, the most numerous and the most important in geochemical history. Without them, the living world would not exist. Thus, according to Darwin's view (1874), the first living beings were not autotrophs, but heterotrophs nourishing themselves on the organic matter produced by photosynthesis. They breathed the oxygen dissolved in the water in which, at a depth of a few centimeters, they were sheltered from the abiotic ultraviolet radiation of the sun.

The liberated oxygen soon formed a layer of ozone above the sea waters which was also fatal for living creatures but had the advantage of preventing the photolysis of organic matter by the near ultraviolet. In the absence of an atmosphere of molecular oxygen, the synthesis of organic matter was brought about by radiations shorter than 2200 Å.

The mitochondria, which are essential living bodies of the cell and are of the size of bacteria, produce, under the influence of the genes, a number of essential substances: The chromoplasts manufacture pigments, such as carotene; the cleoplasts produce lipids; the proteoplasts manufacture protids; the amyloplasts produce starch; and the chloroplasts manufacture chlorophyll.

The appearance of this pigment, chlorophyll, in certain bacteria or in the Cyanophyceae, has completely transformed the living world. The structure of this pigment is based on four pyrrole rings, two of which are joined by a magnesium atom. These pyrrole nuclei had their origin in the pyrogenic syntheses of which we have spoken. It is not a macromolecule (137 atoms) but it has the power, in living matter, of bringing about the photosynthesis of glucides from water and carbon dioxide, and absorbing the 6 quanta energy of red or blue visible light. In other words, in the visible spectrum in which there is much more energy, it effects syntheses which were originally only possible in the far ultraviolet, in the neighborhood of 2000 Å. The fortuitous elaboration of this molecule by the blind play of molecular agitation, valencies, and photochemical action, required a very long time, even on the geological scale, but, once it had appeared, large amounts of oxygen were liberated: The autotrophic plant kingdom appeared with the Cyanophyceae and our present abundant atmosphere of free oxygen was created.

The appearance of chlorophyll finally put an end to the era of ultraviolet photosyntheses, which gave way to photosynthesis in the visible region. The solar spectrum was finally limited to its present value, and all abiotic radiation was arrested by the protective screen of molecular oxygen and ozone. Henceforth, life could appear at the surface of the waters and invade the land masses. Finally, the biosphere was formed.

The later essential stage of organization was the appearance of the cell. But the cell is so complex and contains so many distinct bodies, coordinated by a deoxyribonucleic acid-based nucleus, that it may have taken longer for it to appear than the whole of the recognized period of geological evolution of the living world. The appearance of unicellular creatures, or Protista, marked a definite stage, because the Flagellata gave birth, at one and the same time, to the autotrophic plant kingdom and to the heterotrophic animal kingdom which lived parasitically on it. The single-celled flagellate, such as *Euglena,* presents us with a completed stage in the history of life, since, in it, we find a perfect autotrophic creature, autonomous, mobile, reproducing by division, highly plastic, and possessed of psychism. This psychism must in some way be the result of the molecular dissymmetry, but we shall never understand it because the inquiry is a macroscopic notion that disappears on the molecular and quantic scale.

If such a living creature persisted unchanged throughout the whole of geological time because of its fixed hereditary endowment, it is undoubtedly because it was widely and abundantly distributed originally.

Genetics has proved that evolution came about as the result of rare mutations, caused mainly by the penetrating rays of natural radioactivity and cosmic radiation, acting on countless individuals over thousands of millions of years. During this time, terrestrial radioactivity has decreased to only half, at the most, of its initial value, as a consequence of the progressive exhaustion of the radioactive elements. But the intensity of the cosmic radiation could vary widely during the galactic revolution of the solar system, which occupies a period in the region of 250 megayears. This may be the origin of those "evolutive pushes" recognized by biologists and paleontologists during the geological epochs.

However, mutations occur at random, while evolution has an orderly character and always moves in the direction of increasing complexity, an ever more perfect ascendancy of the living world over the environment and a continuous development of psychism. This is *orthogenesis.* Living creatures are provided with "physical instruments," which are their organs and which have gradually appeared through the blind action of large mutations, which have founded a family of hereditary monstrosities and been preserved by natural selection and the *choice* of the living themselves. Cybernetics has helped to show how the internal feedback that coordinates organisms is controlled by the activity of hormonal secretions and psychism. This feedback, which does not figure in the ordinary, over-simple phenomena of physics, is beginning to show itself in the robots of cybernetics and in Ashby's homeostat. It is striking in the living creature because of its infinite complexity. Thus, the living world possesses a sort of internal finality directing its evolution, just as each individual possesses an internal finality directing its own existence.

The photochemical theory of the origin of life regards it as inevitable that life should appear on a planet at a specific stage of its evolution; a stage that was determined by the surface temperature of its oceans. It follows from this that if life accidentally disappeared from the surface of the earth, the atmosphere of free oxygen created and maintained by the biosphere would disappear with it, together with the ozone, so that the atmosphere would again become transparent in the far ultraviolet and the earth would be uninhabitable. Ultraviolet photosyntheses would soon recur at the surface of the oceanic waters containing ammonium bicarbonate and phosphates. A new life cycle would make its appearance again on the earth. It would still be based on adenosine-phosphoric and deoxyribonucleic acids because of the irreplaceable role of phosphorus.

Among all the planets that constitute the solar system, only our two closest neighbors, Venus and Mars, can be considered from the point of view of cosmic life. The mass of Venus, which is comparable to that of the earth, enables it to retain oceans and an abundant atmosphere. But the planet, which is rich in carbon dioxide and receives only 0.6 of our flux of solar radiation while rotating much more slowly, may be entirely covered by ice. It does not seem that life could have developed on its surface because its atmosphere contains a large amount of carbon dioxide and no trace of free oxygen. The paleovolcanic carbon dioxide has not been retained by the hydrosphere and has not formed, as on earth, important biogenic calcareous formations. Its atmosphere is transparent to the far ultraviolet.

The climatic conditions prevailing on Mars are better known. Because of its small mass, the planet has retained only an insignificant "hydrosphere" which appears as a thin polar cap of hoarfrost. Its very thin atmosphere is rich in carbon dioxide, to the exclusion of all traces of oxygen. Because of its distance from the sun, its climate is harsh. Although the temperature may sometimes rise momentarily above the freezing point of water in its tropical regions, these regions are completely dry and no photosynthesis that would require liquid water could occur, despite the fact that the atmosphere is transparent to the far solar ultraviolet.

The seasonal changes of coloration that have been reported on the planet have not been confirmed by recent work and could be explained as the result of changes in the pattern of saline efflorescence under the influence of the humidity, dessication, and ultraviolet light. The pyrogenous synthesis of heterocyclic organic compounds, which accompanied its original paleovulcanism, have not been followed by photosyntheses in aqueous environment.

But, if we do not find any trace of life in the solar system, this does not signify that, in the universe, life only exists on our planet, and, on the other hand, that planets peopled with living creatures must be legion in the galaxy. The formation of double stars and of planetary systems is the result of stellar interactions occurring in the regions of the universe in which the stellar density is very high, that is to say, at the center of the globular clusters and galactic nuclei. It can be shown that the abundance of the planetary systems is intermediate between that of the red giants and of the double stars, that is to say, it is of the order of hundredths. The galaxy, comprising some two hundred thousand million stars, must contain, when in statistical equilibrium, a thousand million planetary systems. But these are not all similar to ours. Since stellar interactions can occur between stars of any spectral type and of different masses, the resulting planetary systems may be very diverse. They may, for example, contain planets with very eccentric orbits, similar to the orbits of comets, or planets of greater mass than Jupiter, but reduced to the size of tiny hyperdense asteroids; none of which would provide suitable conditions for life. Life requires the fortuitous coincidence of a number of astronomical and climatic conditions, so that in our galaxy life is possible on some millions of planets.

Thus, as the supporters of cosmic panspermia insisted, life and thought are certainly perpetual in the universe, but only in a statistical sense. Life is born independently; it evolves and disappears on every planet on which the physical and chemical conditions for its appearance are momentarily satisfied.

References

1. A. Dauvillier, *Compt. Rend.* **242**, 47 (1956).
2. A. Dauvillier, and E. Desguin, "La genèse de la Vie, phase de l'évolution géochimique." Hermann, Paris (1942).

SUMMARY OF DEFINITIONS, NOMENCLATURE, AND NUMBERS OF PHYSICAL SIGNIFICANCE

Avogadro's number (N): The number of molecules per gram-mole, that is, per 22,410 cm^3 of a gas at standard-temperature (0°C) and standard pressure (760 mm Hg); $N = 6.024 \times 10^{23}$.

Perfect Gas Law: $PV = RT$

where P = pressure in barye units, or in dynes/cm^2,
V = volume in cm^3,
T = absolute temperature in °K, and
R = universal gas constant = 8.314×10^7 ergs/degree.

Boltzmann's constant (K): It is obtained by the dividing of the universal gas constant R by the Avogadro number N; thus, $K = R/N = 1.381 \times 10^{-16}$ ergs/degree $= 3.3 \times 10^{-24}$ calories/degree.

Units of Length: 1 micron (μ) $= 10^{-6}$ m or 10^{-4} cm, 1 Ångstrom (Å) $= 10^{-8}$ cm

Units of Energy: 1 electron-volt (ev) $= 1.6 \times 10^{-12}$ ergs, 1 calorie $= 4.17 \times 10^7$ ergs

If the fundamental relationships of quantum mechanics:

$$W = eV = h\nu = hc/\lambda$$

(where e = the charge on the electron $= 4.802 \times 10^{-10}$ esu or 1.59×10^{-19} coulombs; V = the accelerating potential gradient, volts; h = Planck's constant $= 6.62 \times 10^{-27}$ erg $\times$ sec $= 6.55 \times 10^{-34}$ joules $\times$ sec; c = speed of light in vacuum $= 3 \times 10^{10}$ cm/sec; ν = frequency of the appropriate radiation; and λ = the wavelength of this radiation, in Å) are expressed in terms of λ and V, then one obtains the relationship expressed in practical units:

$$\lambda V = 12{,}350 \quad (\text{Å, volts})$$

Mass of the hydrogen atom: 1.66×10^{-24} g.

Mass of the earth: $M = 5.97 \times 10^{27}$ g.

Surface of the earth: $S = 5.1 \times 10^{18}$ cm^2.

Earth surface not covered by water: 0.28 S.

Solar constant: 2 cal/cm^2 mn^{-1}.

Equatorial earth temperature: +26.3°C

Theoretical average temperature (the earth without an atmosphere): +11.8°C

Actual average earth temperature: +15°C.

Hydrosphere: $M = 1.31 \times 10^{24}$ g; $M/S = 2.56 \times 10^5$ g/cm^2, that is, water to a depth of 2.6 km.

Atmosphere: $M = 5.2 \times 10^{21}$ g, that is, 1,033 g/cm^2, equivalent to 10 m of water.

An atmospheric pressure of 1,033 g/cm^2 corresponds to an air layer 8 km thick, assuming that "standard" conditions (that is, a pressure of 760 mm Hg and 0°C) prevail throughout this layer. This definition of "standard" conditions (STP) applies to all cases where gases are involved.

Carbon Dioxide: 0.4 g/cm^2, that is, 2.2 m of "standard" gas.

Biosphere: $M \sim 2 \times 10^{19}$ g, that is 4 g/cm^2.

Optical microscopy produces enlarged images (either with visible or ultraviolet light) with the aid of lenses or mirrors.

Ultramicroscopy depends on diffraction of light by submicroscopic particles but produces no image of the latter.

Electron microscopy employs electron optics to produce very much enlarged radiographic pictures of ultramicroscopic objects.

I
CHARACTERISTICS OF LIVING MATTER

Terrestrial living matter forms a true surface geological formation to which E. Suess in 1875 gave the name "Biosphere." Chemically, it is composed essentially of water and organic compounds, i.e., belonging to the chemistry of carbon. To a large extent, therefore, it is a tributary of the hydrosphere and the atmosphere, for it draws all its carbon from the carbon dioxide of the air. In this way, it is distinguished from the other, essentially mineral, geological formations that constitute the lithosphere. There are indeed no nonliving organic "species" apart from some waste products of life. Whereas the majority of mineralogical species are derived from silicon chemistry, living species are derived exclusively from the chemistry of carbon, its lower homolog. While minerals are usually produced by exothermic reactions, like silica, for example, which comes from the combustion of silicon ($Si \rightarrow SiO \rightarrow SiO_2$), living matter is, in contrast, endothermic and combustible, and draws its internal energy from a cosmic source, namely, solar light. The energy which gives it life is the result of its controlled oxidation by free atmospheric oxygen. As Le Châtelier wrote (1): "An organic substance, sugar, for example, *tends,* in the presence of the oxygen of the air, to form carbonic acid and water. Neither life, nor any of our laboratory processes can alter this tendency, but they may, both of them, *control* the way in which it comes to fruition."

Numerous criteria for defining life have been proposed: cell structure, nutrition and assimilation, growth, excitability, reproduction, mobility, and psychism. None is entirely characteristic. All granules subject to thermal, molecular Brownian movement show mobility. Something akin to psychism is already apparent in the robots of cybernetics, and the future undoubtedly holds further surprises for us in this direction.

In order to define life, we shall select the criterion of its *energy.* It was first Priestley's, and then Lavoisier's work, that showed the part played by respiration and led to life being defined in terms of energy. It will be an organization on the *molecular level,* drawing the energy actuating it from its controlled oxidation. We shall express it symbolically by the so-called Bayer equation (1870):

$$Q + CO_2 + H_2O \rightleftharpoons CH_2O + O_2$$

or

$$6\,CO_2 + 5\,H_2O \rightleftharpoons C_6(H_2O)_5 + 6\,O_2$$

The endothermic reaction represents the *chlorophyllic photosynthesis* of carbohydrates by the plant kingdom, giving rise, simultaneously, to the atmosphere of free oxygen, while the reverse reaction represents the *respiration* of the living world, the source of all its energy. A man may remain paralyzed and be unable to reproduce: He has nonetheless *lived* between his first and his last breath.

In exceptional cases, oxygen can be replaced by its higher homolog, sulfur. In this way, some sulfobacteria, such as *Beggiatoa alba,* draw their energy from an analogous reaction that can be represented by:

$$CO_2 + 2\,H_2S \rightleftharpoons H{\cdot}COH + H_2O + S_2$$

We shall generally understand, by the term oxidation, all reactions that lead to an atom losing one or several electrons, without necessarily acquiring oxygen or losing hydrogen. One example is the conversion of bivalent iron Fe^{++} to trivalent iron Fe^{+++}.

The energy produced by the oxidation of living matter may appear in the most varied forms: as chemical energy, mechanical energy, thermal energy, electricity, or light.

The work of physiologists has established that life on our level was the resultant of the lives of individual, generally microscopic cells, and that cellular activity itself should be carried back to the molecular level. Cells, isolated in cultures, breathe like microorganisms.

Thus, life is a chemical phenomenon, not on our level, but on the microscopic and *molecular* level. It brings into action dissymmetric stable macromolecules endowed with *genetic continuity. Life is the new form that matter assumes when the asymmetric molecules acquire a high degree of complexity.*

Chemical Composition

Living matter is formed mainly of water—90% for marine creatures and 64.7% for the humans—of the other essential *biogenic* elements, C, N, S, and P, and of numerous heavier elements, Ca, Si, Mg, and Fe, which are present in traces. The immense biogenic importance of phosphorus is due to the fact that, in the Periodic Table, it is the lightest element which is always *pentavalent.* Its composition is very like that of sea water, as statistical chemical analyses have shown; see Table I.

TABLE I
CHEMICAL COMPOSITION OF THE BIOSPHERE AND THE HYDROSPHERE

Biosphere (%)		Hydrosphere (%)	
O	73	O	85.79
C	14	H	10.67
H	9.1	Cl	2.07
N	2.2	Na	1.14
Ca	1	Mg	0.14 or 1.4×10^{-3}
S P Si K	$n \times 10^{-3}$	S Ca K	$n \times 10^{-4}$
Mg Fe Na Cl Al Zn	$n \times 10^{-4}$	C Br N Rb	$n \times 10^{-5}$
Cu Br I Mn	10^{-5}	Si P Fe	10^{-6}
As B F Pb Ti V	10^{-6}	Ni F I B Cu (A)	10^{-7}
Ni Co Sn Mo Cs Rb		Li Cs Sr Ba Co Ti As	10^{-8}
Li La Sr Ba Ce (A)	traces	Mn Al Pb Zn Ag Au	10^{-9}
		Th	$<10^{-10}$
		U	2×10^{-10}
		Ra	2×10^{-15}

Of the forty-eight chemical elements recognized in living matter at the present time, only the first dozen, from oxygen to chlorine, represent 99.98% of its mass. The biosphere is thus very similar to sea water and must, in some way, be derived from it.

The gases produced by the decomposition of living matter are reducing gases and consist mainly of the hydrides, H_2O, CH_4 (or CO_2), NH_3, H_2S, and PH_3. They are analogous to the volcanic gases, H_2O, CH_4 (or CO_2), NH_3, H_2S and HCl. We shall therefore make paleovolcanic exhalation play an important part in the origin of life.

It is possible to make up a *dry gas* having largely the atomic composition of the soft parts of living creatures, e.g., for use in filling the ionization chamber of an X- or γ-ray dosimeter, by adding oxygen, hydrogen, and carbon dioxide to some air. As a matter of fact, the absorption of these radiations by the matter is a strictly atomic phenomenon. Its density is some 0.61 g/l.

The thermal decomposition of living matter that is not in contact with air leads to its carbonization and the release of graphite. This is the origin of the carbonaceous substances scattered in the majority of rocks. The ashes of living creatures are formed mainly of the oxides of the elements Ca, Si, and Fe.

The chlorophyll pigment of plants contains 3×10^{-3} of magnesium.

The respiratory pigments of invertebrates contain some copper or iron. The elements Mg, Mn, Cu, and Zn are contained in enzymes and diastases, the catalysts of living matter, in vitamins, borrowed from the plant kingdom, and in hormones, secreted by the living creature itself.

The works of J.-B. Dumas, J. Boussingault, A. Gautier, and G. Bertrand have shown that a great many elements, the oligo-elements, although present in traces, are useful or indispensable to life.

Plants contain constant trace amounts (10^{-5}) of Si, Cl, Na, Mn, and Al, and variable trace amounts of I, Br, F, As, B, Rb, Li, Sr, Ba, Zn, Cu, Co, V, and Ce. The elements I, B, Fe, Mn, Cu, Zn, Mo, Mg, and Co are indispensable to animals. Fertilizers contain mainly N, P, S, K, and Ca. It is because lavas and volcanic ashes come from the molten magma, in which *all* the chemical elements are present and intimately mixed, that volcanic soils are so fertile. Thermal springs seem to owe their properties to these elements. Sea water contains all the oligo-elements.

Organisms derive their elements from other organisms, from the dust of the earth and air, and, above all, from aqueous solutions. It is invariably the same atoms that are utilized in the formation of living creatures which emerge, for a moment, from the molecular chaos and soon return. The mass of the biosphere is extremely small compared to the mass of living matter that has existed.

It is notable that the biogenic elements are among the lightest in Mendeleyev's Periodic Table. Indeed, it is essential that they be able to give rise to soluble, volatile gaseous compounds. The compounds formed by their higher homologs are, in contrast, solid, refractory, poorly soluble in water, and able to form the minerals of the lithosphere only. Meanwhile, among the halogens, all of which are volatile, only bromine and iodine are utilized in living structures, while fluorine and chlorine are almost completely excluded.

The role of water is fundamental. Life could not have been conceived without water in the liquid state. It is the only inorganic liquid in nature and the solvent of numerous salts. Its high dielectric constant permits reactions between ions and between colloids. Its specific heat is greater than that of all neutral organic liquids. Its high heat of vaporization acts as an indispensable thermal "energy store" at the surface of the globe. The latter was, in great part, covered by sea water, and Goldschmidt laid stress on the fact that each square centimeter was covered, on the average, by 273 liters of water. The marine environment is just alkaline, containing a slight excess of OH^- ions. It contains 10^{-9}% of H^+ ions, that is to say, it has a pH of 8.17. Life can only be maintained, in the marine environment, between the limits pH = 5 and pH = 10. The pH of living creatures is 7.5. That it is remarkably close to that of sea water is no coincidence. After death it falls to 5. Life could not appear and be maintained on a planet where the water was always in solid or gaseous form, as on Mars and Venus.

The role of water is shown clearly by the fact that life is put into suspension by *desiccation* which, like refrigeration, is a conservation process that does not destroy structure, as chemical or thermal sterilization does.

The solubilities in water of carbon dioxide, nitrogen, and oxygen play an essential role in the maintenance of marine life, in its nutrition and respiration. If the oceans had been covered by a film of mineral oils of volcanic origin, for example, life would never have appeared on the earth.

Although formed mainly of water, living matter is not soluble, on account of its physical state. It forms *gels* and semiwettable membranes. *Plastics* are the artificial organic substances closest to it.

One may say that life is so aquatic a phenomenon that the higher living creatures, on leaving the marine environment, have "borne" it away with them in their plasma (E. Desguin). Sea water *passes through* marine creatures surprisingly quickly—as through a filter—giving up its nutritive elements to them.

"L'azote" is an indispensable constituent of life, as it is of the majority of explosives. It is therefore very badly named and it would be more suitable to call it *nitrogen*, as is done abroad. Life is no more conceivable without its higher homolog phosphorus, an indispensable constituent of nucleic acids. Apatite, $P_2O_5 \cdot CaO$, which is found in almost all rocks in the proportion of some thousandths, is probably the source of this element. The marine environment contains the PO_4^{---} ion. The homolog of phosphorus, arsenic, is encountered only in traces, and is toxic in greater amounts; the same applies to antimony.

Living matter, produced mainly from the gases H_2O, CO_2, and NH_3, returns to the gaseous state through decomposition. As examples, we may cite methane, CH_4, which comprises marsh gas and firedamp, CO_2, H_2, H_2O, H_2S, N_2, and NH_3, as well as numerous odorous volatile organic substances. Marsh gas is produced by the reaction:

$$C_6(H_2O)_5 + H_2O \rightarrow 3\ CO_2 + 3\ CH_4$$

Phosphoretted hydrogen phosphide appears in the "ignis fatuus."

A number of authors have thought, naively, that with other chemical elements, other forms of life would be possible, in physicochemical conditions different from ours. Preyer (1875) already had this idea when he imagined his "pyrozoairia," legendary creatures that could have lived in fire. These lucubrations reveal nothing more than a tiresome failure to appreciate chemistry. When, for example, silica replaces phosphoric acid in ribonucleic acid, a fatal disease, silicosis, is the result. Silicon could not, by any means, replace carbon. Unlike carbon, silicon is linked to another silicon atom only through the intermediary of an oxygen "bridge." This is the siloxane radical, Si—O—Si, to which methyl, ethyl, and phenyl radicals can be attached. Its oxide, instead of being volatile and soluble, is refractory and insoluble. It has given rise to a host of minerals, but its "organic" chemistry remains very rudimentary. Apart from the silicones, which are stable, only a small number of saturated and unsaturated hydrogen silicides, siliciformic and silicioxalic acids are known; they are all very unstable. Silicon has hardly been utilized by the living world except in forming the silica skeleton of the Diatoms and Radiolaria. The universality of the chemical elements requires no further demonstration. There is only one carbon chemistry and we may be sure that the living creatures inhabiting other heavenly bodies make use of the same organic chemistry as we do, although with a different isotope composition. The isotope composition is uniform in the solar system but different for each star.

The Chemistry of Carbon

The lightest of the tetravalent elements, carbon, is unique in the Periodic Table in that it forms with oxygen and nitrogen a multitude of hydrides and ternary and quaternary compounds. Nevertheless, not all the innumerable theoretically possible combinations, $H_nC_mO_pN_q$, are realized on account of the role played by valency in molecular architecture. Generally, the condition is: $n > m > p > q$. Carbon already forms C_2, C_3 . . . C_n molecules in the stars, where it forms cosmic soot clouds. It may give rise to infinite molecules in the crystals of graphite and diamond. Situated on the line dividing the electropositive from the electronegative elements, it may combine with the elements of both categories. It is indispensable in forming differentiated molecular skeletons, in long straight chains, or in cyclic Kekulé nuclei. It is possible to say that organic chemistry and, consequently, living matter are the result of the specific properties of the carbon atom.

Finally, the essential fact, the *asymmetric carbon* is the principal source of the optically active compounds characteristic of living matter, as Pasteur showed.

Before this, from the time of Lémery (1675), all compounds extracted from the plant and animal kingdoms were included in organic chemistry. They were then considered as having a special "essence." It was later admitted, with Berzelius, that they were differentiated by the fact that they could arise only in organisms and under the influence of a hidden, so-called "vital," force. In 1828, the synthesis of urea by F. Wöhler (1800–1882) showed that there was no such difference in the nature of inorganic and organic compounds.

A whole series of remarkable syntheses were due to Marcelin Berthelot (1827–1907) who carried out the synthesis of acetylene (1864) by combining elements in the electric arc, as a starting point. His pyrogenic condensation led to numerous benzene ring C_6H_6 carbides (1886), to naphthalene, and to anthracene. But as far back as 1853, he had already succeeded in synthesizing fats, ethyl alcohol, and formic acid.

Living matter utilizes some fairly simple ternary compounds (H, C, and O), such as carbohydrates or glucides (sugars, starch) and fats, as well as quaternary compounds (H, C, O, and N) forming the polarized chain *amino acids,* in which one end is acid $—CO_2H$ and $—NH_2$; amination occurs in the α position. Some heavier elements, S, P, and Fe, act as linkages in the formation of *proteins,* the bases of life. The cyclic nuclei of aromatic chemistry play an important role in the sterols. The halogenated, nitrated, and acetylenic derivatives, on the other hand, are toxic. Alkaloids, vegetable oils, and resins, which are frequently ternary compounds and are excreted by the plants, seem to be waste matter which no living creature can assimilate.

Moureu and Dufraisse demonstrated the role of the phenols (tannins, creosote, guaiacol) in plants: They are agents that protect against a too rapid oxidation and permit life to be slowed down.

The synthesis of polypeptides, which play an essential role in living matter, was achieved by E. Fischer from glycine, $NH_2 \cdot CH_2 \cdot CO_2H$, the second term of the amino acid series. In 1903, he synthesized a polypeptide with eighteen amino acids and a molecular weight of 1268.

Molecular Dissymmetry

In 1812, Biot discovered the rotatory molecular power of a kerosene oil and then of a large number of organic substances, and it was soon realized that organic substances endowed with rotatory power were characteristic of living matter. It was thought that it would remain impossible to synthesize them. Pasteur, in 1848, linked optical activity to crystal dissymmetry and showed molecular dissymmetry to be characteristic of living matter. Le Bel, in 1874, and Van't Hoff, in 1875 constructed the theory of the asymmetric carbon. Finally, in 1873, Jungfleisch achieved the synthesis of mixed *dextro-* and *levo*-tartaric acids from ethylene, and in 1888, Ladenburg, synthesized cicutine (conicine, conine), the alkaloid from hemlock.

In exceptional cases, optical activity may be due to the pentavalent nitrogen. The hexavalent cobalt and platinum of Werner's complexes can also confer molecular rotatory power, which is, therefore, not an exclusive characteristic of living matter, or even of organic substances.

The theory of the asymmetric carbon greatly increased the number of isomers. Thus, Blair and Henze have calculated that, for the formula $C_{20}H_{41}(OH)$, there are 82,299,275 possible alcohols, of which 82,287,516 are active and 11,759 are inactive.

Macromolecules

In 1861, Graham discovered a new state of matter, seemingly intermediate between inorganic molecules and living matter. This was the *colloidal* state, which stimulated a great deal of research. It was soon recognized that these opalescent pseudosolutions were suspensions of very large molecules, even rather of fine particles, stirred by the Brownian movement. In sols, the maximum size of the granules does not exceed 0.1 μ, whereas in suspensions it is over 1 μ. The granules are surrounded by a double electrical layer and generally carry a charge. The principle of electrophoresis depends on this. If the particles are a micron in size, the width of the double layer is of the order of 100 Å.

Van't Hoff explained the osmotic pressure of these solutions by the gas laws, and Perrin showed that emulsions of microscopic granules also obeyed these laws within wide limits.

The molecules of inorganic chemistry attain their highest molecular weight with Werner's complexes. We can cite, for example, chloropurpureocobalt phosphate $[Co(NH_3)_5Cl]_2P_2O_7$, which contains fifty-three atoms and has a molecular mass of 533. We shall see that the molecules of organic chemistry can be incomparably more massive.

The study of inorganic colloids, such as the powdered metals produced in the electric arc (Bredig) and studied in the ultramicroscope, showed that their active chemical properties were due to their extreme fineness, that is to say, to their very large surface area. Even earlier, catalysts were known, such as unpurified platinum or palladium sponge, which combine the elements of the detonating gas at ordinary temperature. Like the fermentation that turns wine into vinegar, colloidal platinum decomposes oxygenated water and oxidizes alcohol to acetic acid. Its properties are destroyed by a rise in temperature, as well as by antiseptics and poisons. In 1904, Noel Bernard called these metal "ferments" "inorganic diastases."

Pasteur's theory that living cellular organisms caused these fermentations was soon replaced by the theory that molecular diastases or enzymes produced slow oxidations such as fermentations and respirations.

In 1920, Staudinger gave the name "macromolecule" to the huge organic structures forming sols and gels, and having new properties, unknown for molecules such as glycerin or urea. These formations were studied by numerous methods: ultrafiltration, ultracentrifugation, osmometry, chromatography or capillary analysis, ultraviolet and infrared spectra, X-ray diffraction, and others. Their size, mass, and structure, which were beyond the range of the ultramicroscope, were established by the electron microscope. Their synthesis has led to the artificial fiber and plastic industry, these substances resulting from the polymerization of simple molecules, such as:

polystyrene $(C_8H_8)_n$
Teflon $(—CF_2—CF_2—)_n$
Plexiglas $(=CH_3—\underset{\underset{CH_2}{\|}}{C}—CO_2CH_3)_n$
polyethylene $(—CH_2—CH_2—)_n$
polyvinyl chloride $(—CH_2—CHCl—)_n$
nylon $[CO—(CH_2)_4—CO—NH—(CH_2)_6—NH—]_n$.

As binary colloids (H,C) we may cite rubber and gutta-percha; as ternary colloids (H, C, O), starch, glycogen, dextrin, agar, tannin, gums, and mucilages; as complex colloids (containing nitrogen, sulfur, phosphorus, and iron, in addition to H, C, O) albumin, fibrin, casein, gelatin, collagen, and proteins.

In traversing microscopic and ultramicroscopic suspensions there is *continuity* from the precipitates with granules visible to the naked eye up to the high polymers (2).

The macromolecules characteristic of living matter are the result of the combination of *proteins* and *nucleic acids.*

Proteins are macromolecules whose mass lies between 3×10^4 and 3×10^6, and whose skeleton is a long linear chain produced by the joining together of amino acids:

$$NH_2—\underset{R_1}{\underset{|}{CH}}—CO_2H \ldots NH_2—\underset{R_n}{\underset{|}{CH}}—CO_2H, \text{ etc.}$$

with elimination of a molecule of water.

These heteropolar molecules are thus aligned like small magnets, by a carboxyimide bond, forming polypeptide chains:

$$NH_2—\underset{R_1}{\underset{|}{CH}}—CO\left[—NH—\underset{R_2}{\underset{|}{CH}}—CO—\right]_n—NH—\underset{R_p}{\underset{|}{CH}}—CO_2H.$$

W. T. Astbury showed that the amino acids were not always inserted at random into the chain, but that they had a *periodic* arrangement. Thus, in silk fibroin (see Fig. I-1) glycine and alanine are repeated regularly:

$$\begin{array}{c}\mathrm{H}\\ |\\ \mathrm{NH_2{-}C{-}CO_2H}\\ |\\ \mathrm{H}\end{array} \quad \text{and} \quad \begin{array}{c}\mathrm{H}\\ |\\ \mathrm{NH_2{-}C{-}CO_2H}\\ |\\ \mathrm{CH_3}\end{array}$$

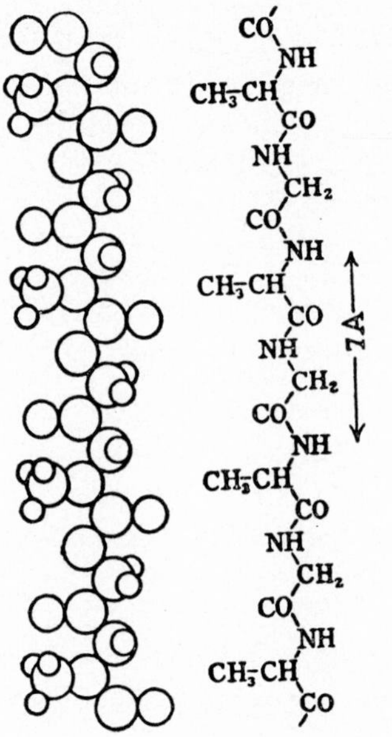

Fig. I-1. Structure of silk fibroin (after W. T. Astbury).

This substance, which possesses a central asymmetric carbon atom, is endowed with rotatory power. Throughout the entire biosphere, only *levo*rotatory alanine is found. The natural *dextro* amino acids are extremely rare.

The amino acids may be aliphatic, like alanine; aromatic, like tyrosine; acids, like glutamic acid; amides, like asparagine; basic, like lysine; cyclic, like proline; or contain sulfur, like cystine.

Meyer and Mark have shown that keratin, a constituent of hair, feathers, nails, and horns, has a similar structure. It is a thermal and electrical insulator, hygrometric and elastic. This elasticity is explained by the folding and unfolding of the long chains. When the bonds between the carbon and nitrogen atoms are broken, electrons are liberated. This explains the negative electrification caused by cutting hair or stroking a cat's fur.

It is remarkable that the amino acids which are utilized in the formation of living matter are only about twenty in number. The most important are leucine, glutamic acid, arginine, proline, cystine (which contains sulfur), alanine, and tyrosine. Nevertheless, innumerable structural combinations are possible. In the crystalline state they are *phosphorescent* in ultraviolet light, by which they are decomposed. This is one method of analyzing these compounds.

The first protein in which the order of the amino acids was completely determined was insulin (5733), the pancreatic hormone controlling the glucose concentration of the blood. It comprises two chains of twenty-one and thirty amino acids, linked by disulfide bridges. The amino acids are distributed *at random* but their arrangement determines the unique biological activity of the hormone (3).

The *nucleic acids* are macromolecules whose basic element is the *nucleotide*, formed by the combination of a phosphoric acid molecule, a sugar, such as ribose or deoxyribose, and a purine or pyrimidine base. Phosphorus pentoxide, P_2O_5, gives, with one molecule of water, metaphosphorc acid, HPO_3; with 2 H_2O, pyrophosphoric acid, $H_4P_2O_7$; and with 3 H_2O, orthophosphoric acid, H_3PO_4. It is known that the latter can give mono- or dialcoholic esters such as:

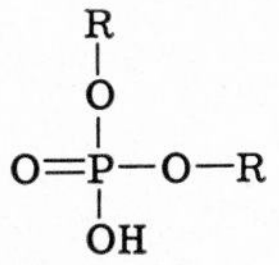

```
        R
        |
        O
        |
    O=P—O—R
        |
        OH
```

The enzymes, which are catalysts, thus consist of a protein linked to a nucleotide. Adenylic acid, the nucleotide of the dehydrogenase of muscle and yeast, consists of adenine, ribose, $C_5H_{10}O_5$, and phosphoric acid.

Adenosine triphosphate (ATP), which is a constituent of ribonucleic acid, is a combination of purine and ribose which utilizes the unique phenomenon of the abundant free energy of pentavalent phosphorus:

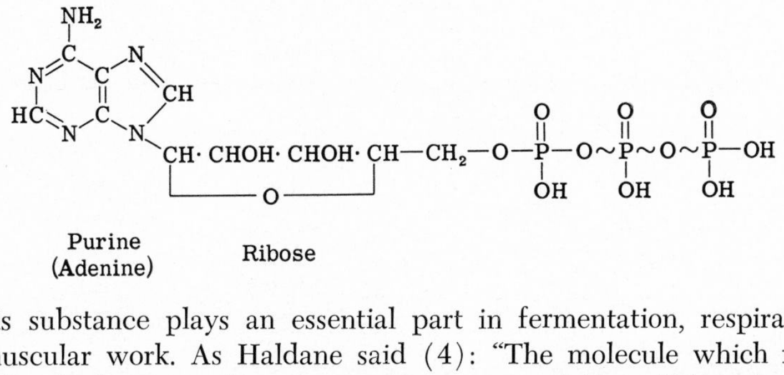

```
      NH2
      |
      C     N
   N//  \C/  \\
   |     ||    CH
   HC\\ /C\   /          O        O        O
      N     N            ||       ||       ||
             \CH· CHOH· CHOH· CH—CH2—O—P—O~P~O~P—OH
              |                |          |       |       |
              |_______O________|          OH      OH      OH

   Purine
  (Adenine)          Ribose
```

This substance plays an essential part in fermentation, respiration, and muscular work. As Haldane said (4): "The molecule which most resembles the living is adenosine triphosphoric acid. Two ATP molecules may each give up a phosphorus radical to a glucose molecule which is breaking down. The products are, first, diphosphoglyceric acid, then phosphopyruvic acid, which will then give up their phosphorus radicals to 4 molecules of adenosine diphosphoric acid (ADP), giving 4 molecules of ATP. These phenomena occur in fermenting yeast and in our muscles. ATP may thus reform its lost parts: one can say that it possesses a *metabolism.* In a still more surprising way, a phenomenon analogous to a sexual process occurs when a large amount of labile phosphorus has been lost (the 3rd phosphorus radical of ATP). Two molecules of ADP may then give rise to one molecule of ATP and one molecule of adenylic acid. The latter is generally rapidly broken down by deamination, like a polar sphere, if we insist on pursuing the comparison."

The nucleic acids may be associated through their phosphoric acid, and may form, for example, tetranucleotides. Zymo- or ribonucleic acid (RNA) has the composition:

```
   OH
   |
O=P—O—Ribose—Uracil
   |        |
   OH       O
            |
         O=P—O—Ribose—Adenine
            |        |
            OH       O
                     |
                  O=P—O—Ribose—Cytosine
                     |        |
                     OH       O
                              |
                           O=P—O—Ribose—Guanine
                              |
                              OH
```

It is characteristic of plant viruses and bacteria.

In another essential nucleic acid, thymo- or deoxyribonucleic acid (DNA), the ribose is replaced by deoxyribose. It is characteristic of genes, animal viruses, and bacteriophages.

Watson and Crick (5) have suggested that, in the gene, virus, and phage, two helical DNA chains, whose atom sequences run in opposite directions, are coiled around a common axis at a rate of 1000 turns for a molecular weight of 6,700,000.

In the phage, this thread is surrounded by a protein envelope. In chromosome replication the two chains separate longitudinally by rupture of the hydrogen bonds linking the nucleotide bases.

It has been recognized that deoxyribonucleic acid has antiferromagnetic properties. Douzon *et al.* (6) have found it to possess electrical and ferroelectrical properties. The nature of these discoveries is such as to throw light on the problem of cell division.

In deoxyribonucleic acid, the purine base may be adenine or guanine and the pyrimidine base, cytosine or thymine. Figure I-2 shows the

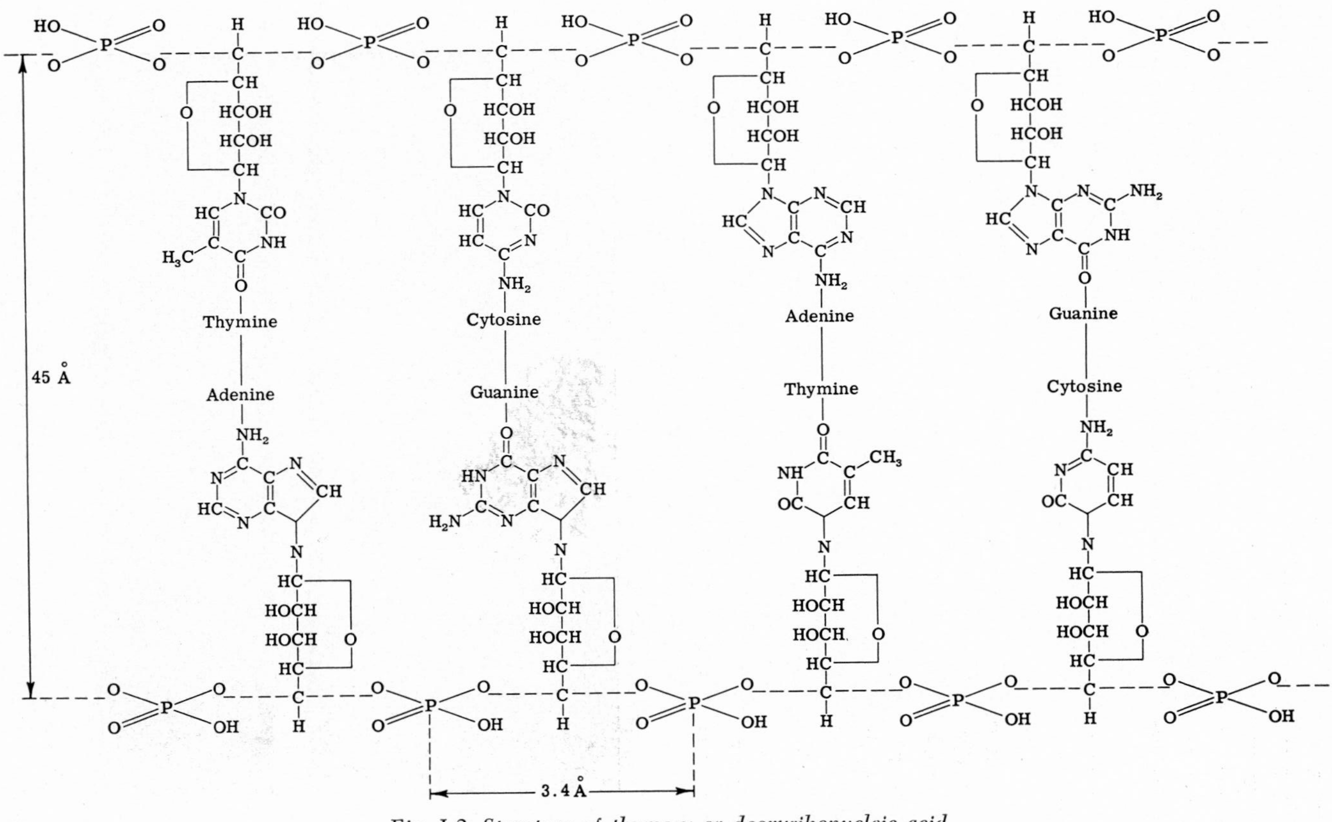

Fig. I-2. Structure of thymo-γ or deoxyribonucleic acid.

structure of thymo- or deoxyribonucleic acid (DNA) according to Watson and Crick (5). All the possible combinations between twenty or fifty amino acids are far from being realized, on account of the existence of protein subunits. With a link having a molecular mass of the order of 300, a chain of 245 links corresponds to 73,500. The maximum is of the order of 500,000. Ochoa and Kornberg succeeded in biosynthesizing ribonucleic and deoxyribonucleic acids.

According to W. M. Stanley, the chromosome structure would be written by means of the four molecules, adenine, guanine, thymine, and cytosine.

This chemistry already provides an image of life. In it we see self-reproducing macromolecular *races* coming together and living symbiotically.

Some macromolecules are able to crystallize. The characteristic crystals of hemoglobin ($M = 68{,}000$), prismatic or tetrahedral, depending on their origin, have been known for a long time (F. L. Hunefield, 1840).

Pasteur had shown, in establishing bacteriology, that alongside the bacteria, which were stopped by filter candles of unglazed porcelain, there existed *filterable viruses* which were able to pass through. They also passed through the pores of collodion filters. The porosity of these

Fig. I-3. Crystals of yellow mosaic virus of turnip. (*Negative of R. Markham, Agricultural Research Council, Molteno Institute, Cambridge; from P. Morand* in *"Aux confins de la vie," Masson, Paris, 1955.*)

candles varies from 12 to 0.3 μ, that is to say, these were submicroscopic viruses. The only filterable virus visible in the ultramicroscope (10 mμ) was that causing contagious bovine pleuropneumonia. These ultraviruses produce infectious diseases such as influenza, measles, poliomyelitis, smallpox, yellow fever, rabies, foot and mouth disease, rinderpest or swine fever, fowl sarcoma, and silkworm grasserie. The virus causing tomato bushy stunt is remarkable in the sense that F. C. Bawden and N. W. Pirie showed, in 1938, that it crystallized in rhombododecahedra from a cubic seed. Figure I-3 shows the crystals of turnip yellow "mosaic" virus. The virus causing the "mosaic" disease of tobacco, discovered by Iwanovsky in 1892, and detectable by mottling of the leaves, can also be crystallized, as Stanley showed in 1935. From 1 kg of tobacco leaves, it is possible to extract about 1 g of a substance that crystallizes in fine needles, 20 to 30 μ in length, and has a molecular mass of 25×10^6. Organic analysis gives O = 73.3, C = 50, N = 16.5, H = 6.8, and ashes = 10^{-3}

In 1941, Stanley succeeded in obtaining an electron micrograph of this virus, where it appeared as rods measuring 15 by 150 mμ. This was the first picture of a macromolecule that was obtained (Fig. I-4).

A remarkable experiment of Steere and Williams of the Virus Laboratory at Berkeley, consisted in extracting, by freezing and micromanipulation, a 20-μ mosaic crystal from a tobacco leaf cell. The latter, when placed in water, is "dissolved" into particles of virus.

X-ray analysis showed that the protein thread was wound in helical form around a nucleic acid rod, and the helix consisted of sixteen subunits of mass 18,000 per turn. Nixon and Woods (7) succeeded in obtaining the end-on image, in the electron microscope, an image on which the subunits, which measure 20 Å, could be counted directly.

Figure I-5 reproduces an electron micrograph of a tobacco necrosis virus crystal at such a magnification (57,000) that the virus particles are evident *in situ* in the lattice.

This crystallizable nucleoprotein, when inoculated in a dose of 0.01 mg, is capable of transmitting the disease to twenty-nine plant species. It multiplies in the host cell at the expense of the host cell's proteins, but remains inert *in vitro*. The protein viruses of plant and animal diseases are, like pathogenic bacteria, destroyed by sterilization, antiseptics, ultraviolet rays, and X-rays. In effecting the synthesis of their own molecules, they assimilate, but, as Bronfenbrenner and Reichert (8) showed in 1936, *they do not respire,* and, therefore, are not living.

As F. C. Bawden wrote: "It is not the fact that viruses form solid crystals which separates them so profoundly from organisms, for it would be possible to conceive of small organisms arranging themselves

Fig. I-4. Rods of tobacco mosaic virus. (Trub electron micrograph, Tauber & Co., Zurich.)

in space with a 2- or even a 3-dimensional regularity. The real differences rest rather on the chemical simplicity of the viruses and on the regularity of their internal structure. All known *organisms* contain salts, sugars, and numerous substances other than proteins, but viruses seem to be simple solid masses of nucleoproteins which do not contain appreciable amounts of either water or other diffusible substance. In addition, X-ray measurements show that the subunits of which the virus particles are constructed, are arranged in the same manner and with the same degree of regularity and complexity as in other well-defined proteins, such as keratine of feathers."

The multiplication of viruses shows that, beyond a sufficient size, these macromolecules may acquire a complexity which confers on them

Fig. I-5. Electron micrograph of a tobacco necrosis virus crystal. (Magnification: ×57,000). (After R. W. G. Wyckoff and L. W. Labau.)

a characteristic that was previously considered purely biological. This fact ought not to cause surprise, for the large tobacco mosaic virus already attains the size of the smallest bacteria, such as *Rickettsia exanthematotyphi.*

As W. M. Stanley remarked: "The probability of formation of a unit of matter by *chemical combination* decreases as its complexity increases, because its combination depends on the hazard of encounters and of the component units. On the other hand, the probability of formation by reproduction of units of matter having biological characteristics increases as the complexity increases, because the voluminous micellae offer a shelter to the compounds which penetrate into their interior, transform them and assimilate them. Thus they grow, first in size, then in number."

The duplication of the virus macromolecules is not achieved by fission, like cell division, but by duplication. Jordan has proposed, on theoretical grounds, the image of the printing block and of its replica, but if this were so, the two structures would be mirror images, like the right and left hands, instead of being identical.

The sexual characteristics, which are already evident in viruses, show them to consist of 30 to 50 genes capable of multiplying independently. The mass of the virus, which is of the order of 10^{-17} g, is effectively a hundred times greater than that of the gene, 10^{-19} g, which is a macromolecule made up of some 10^4 biogenic atoms.

In 1955, Fraenkel-Conrat and Williams (9) succeeded, at Berkeley, in separating tobacco mosaic virus into protein and nucleic acid, both inactive, and in reproducing a feebly active virus by reuniting them again. All these works confirm that it is not living matter that we are dealing with, yet it is remarkable that the basis of the heredity of the living world, the gene, is itself nonliving.

These phenomena throw some light on a particular disease of insects, polyhedral disease. The substance of the insect crystallizes and causes death. The virus responsible has converted the living organism into its proper, nonliving, crystalline substance. Living matter could not be conceived in the solid state, any more than in the gaseous state.

Like the ultraviruses, the bacteriophages, discovered by d'Hérelle in 1915, are macromolecules that produce colonies in the bacterium which then dies and breaks up. In half an hour, the phage has given birth to 20 descendants which, however, utilize only a small fraction of the bacterial substance.

The electron microscope showed (Fig. I-6), by means of a 30,000 diameter enlargement, that bacteriophages were in the shape of ovoid spheres of about $1/15\ \mu$, that terminated in a caudal "cilium" of the same length which, in actual fact, was a microsyringe for injection. They are seen moving toward the bacteria, as if under the influence of a directing "field." Only one species is known in the intestine of all vertebrates and invertebrates. The phage is likewise inert *in vitro* and multiplies only in a living host.

Hershey and Chase showed (see Fig. I-7), by using labeled elements, how the phage injected its internal filament P, which is responsible for its intrabacterial reproduction at the expense of the proteins of the bacterium.

The "large" bacteriophages are more sensitive to ultraviolet rays than the "small." There is an analogy here with the sensitivity of a photographic emulsion or luminescent crystalline screen as a function of grain size, which is explained by the quantum theory.

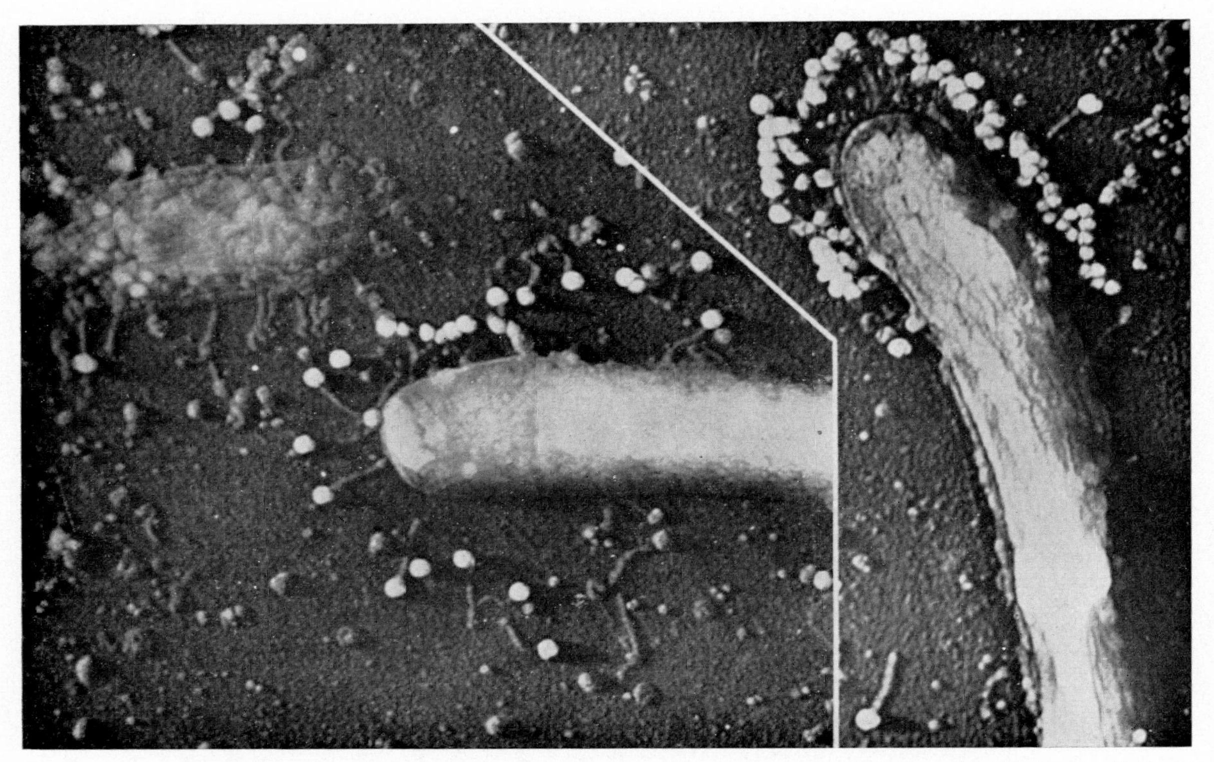

Fig. I-6. Bacteriophages attacking a bacterium. (*Negatives by G. Penso, Institut Superieur de la Santé, Rome; from P. Morand in "Aux confins de la vie," Masson, Paris.*)

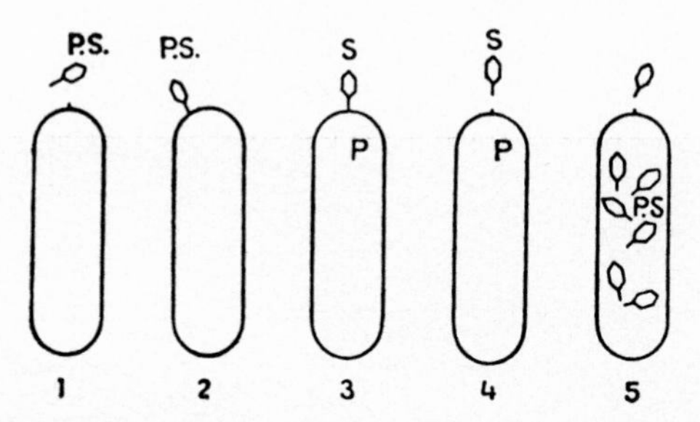

Fig. I-7. Diagram showing a phage injecting a bacterium. (*P* = *protein;* S = *phage envelope.*) (*After Hershey and Chase.*)

Living Creatures

The simplest living creatures, such as bacteria, are of a completely different order of complexity to viruses and phages. The living creature is characterized—omitting respiratory metabolism—by its *organization* and its *psychism,* which are consequences of its complexity. It comprises *organs* which are *instruments,* such as circulation "pumps," *cilia* or *flagella,* for mobility, touch, or grasping. It is characterized by a central body containing Osborn's "hereditary chromatin" (because of its capacity to take up stains), which consists of *chromosomes,* formed of numerous *genes,* bathed in the cytoplasm that is a hyaline liquid in perpetual circulation in the living state. The respiratory or chlorophyll pigments are contained in the cytoplasm. This is so in the case of the Cyanophyceae and ensures that matter and energy are exchanged with the external environment. Bacteria reproduce by fission, which provides identical copies. The genes have the same chemical nature as animal viruses and phages. They contain the heredity, or genetic continuity, of the living. This is why the injection of deoxyribonucleic acid, extracted from one particular bacterial family, into another close family, may result in the appearance in the latter of characteristics peculiar to the first. Benoit has extended these results to the higher animals.

These generally found proteins testify to the unity of living matter and to their own unique origin. The fact that primitive species still exist and have remained unchanged for five hundred million years, while passing through a much larger number of generations, establishes the molecular character of the genes. Molecules and crystals alone show a similar perenniality because of their invariable structure.

This means that an abyss of complexity exists between the virus and the living organism. The size of the living organism is of a greater order of magnitude. Although the mass of an oleic acid molecule, $C_{18}H_{24}O_2$, 2.2 mμ in length, is only 4.6×10^{-22} g ($M = 282$), that of the tobacco mosaic virus macromolecule reaches 4×10^{-17} g ($M = 40 \times 10^6$), and that of the smallest bacteria is some 10^5 times greater. These ratios are shown on the logarithmic scale in Fig. I-8.

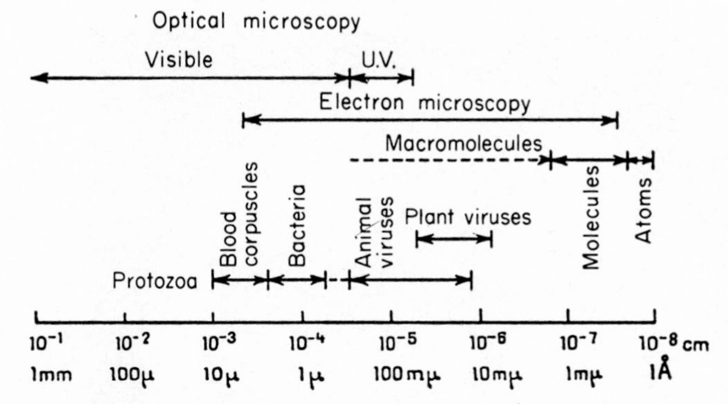

Fig. I-8. Logarithmic scale representing the comparative dimensions of atoms, molecules, macromolecules, viruses, bacteria, and unicellular organisms.

Microorganisms live in a world totally different from our macroscopic world, since, for them, molecular forces are much more important than hydrodynamic forces.

Figure I-9 shows, at two different magnifications, a common bacterium, *Proteus vulgaris.* The first (Pasteur Institute) is observed in the normal microscope ($\times$ 1850 diameters), and the second is observed in the electron microscope, at a magnification of 62,000. This second negative is of extreme interest in that it compares the multiple bacterial flagella with the rods of tobacco mosaic virus. Although the diameter of the latter is 15 mμ, the flagella do not exceed 12 mμ, that is 120 Å. They represent the finest structure discernible in living creatures.

The X-ray diffraction studies of Astbury and Weibull (10) have shown, for these flagella, the same diffused rings as those of keratine. They are bundles of long protein molecules. In the illustration we have represented diagrammatically, at a magnification of 3×10^7, what the molecular aspect of one part of a flagellum would be if it were observable on this scale. The lateral bonds, which are not shown, between the longitudinal chains, are made by opening the double bonds of the CO carbonyl groups. Moreover, other equally fine natural structures are known. So it is that one very small spider, Meioneta, spins fibroin threads of 50 to 200 Å, the finest having only one section of 7 molecular chains.

The flagella of unicellular organisms are considerably larger and more complex. Figure I-10 shows the flagellate, *Euglena gracilis* ($10 \times 50\ \mu$). The single short flagellum has a diameter in the region of

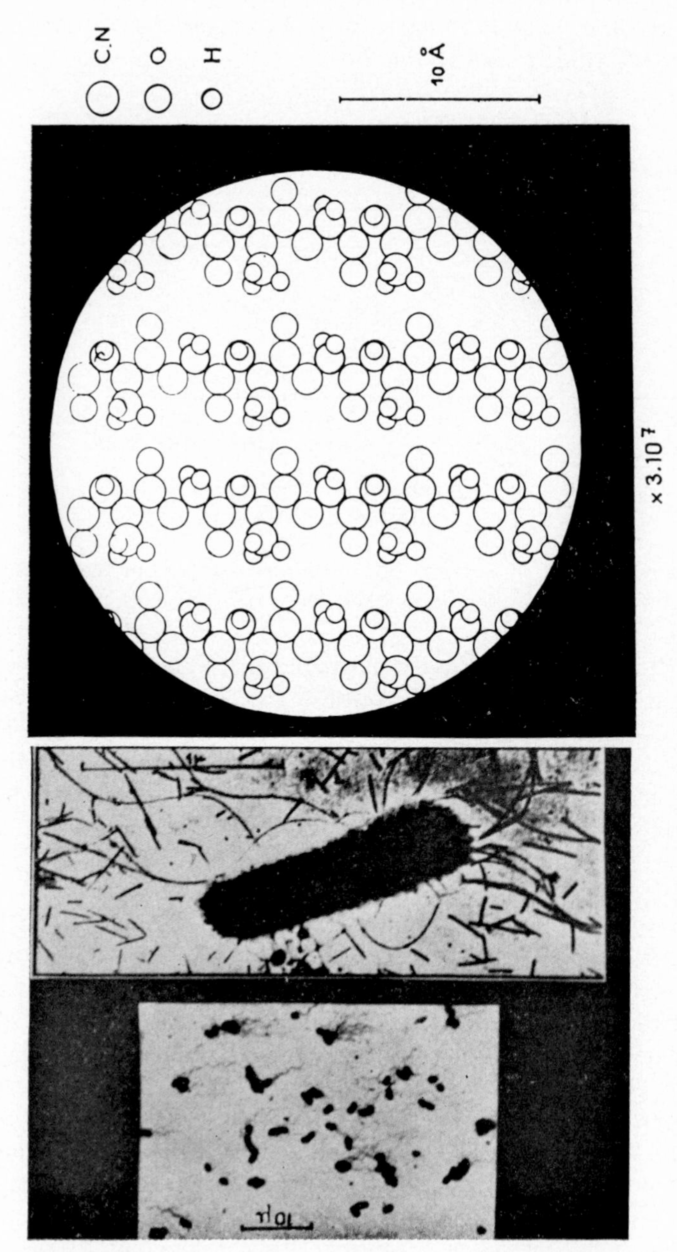

Fig. I-9. Proteus vulgaris observed in the microscope (×1850) and in the electron microscope (×62,000) in the presence of tobacco mosaic virus. The diagram shows the molecular structure of quarter of the section of a flagellum if it were enlarged 3×10^7 times.

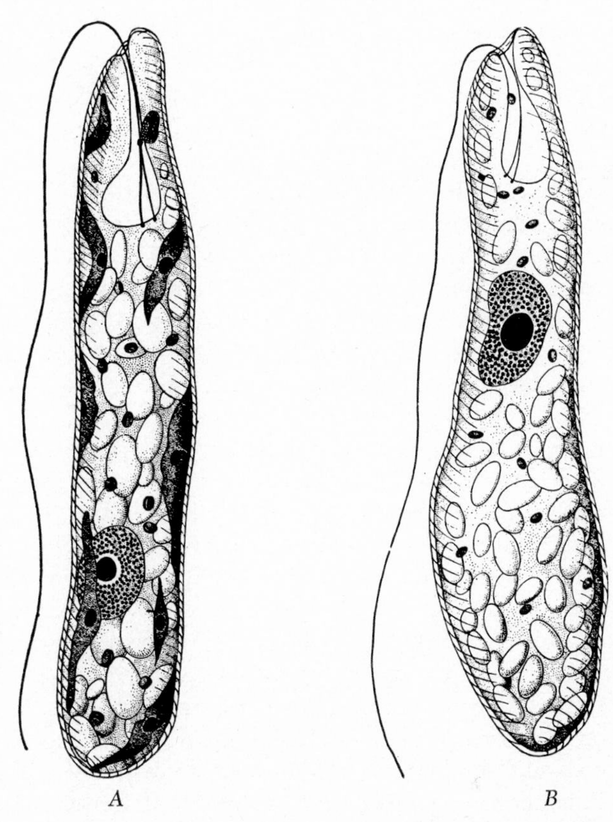

Fig. I-10. Complex structure of a unicellular organism with flagellum. A. Euglena gracilis, with dictyosomes, plasts and pyrenoids, paramylon, and stigma. B. Euglena gracilis var longa, dictyosomes (From A. Holande in "Traité de Zoologie," vol. I, p. 1, Masson, Paris.)

1 μ. Some of the numerous intracellular bodies can be seen in the cytoplasm.

Thus, the phage macromolecule, the bacterium, and the single-celled organism are constructed in a similar structural plan, but with steeply increasing dimensions and complexity. But there is a greater gap between the bacterium and the unicellular organism than between the latter and the multicellular organism. Man is made up of some 10^{27} atoms forming 17×10^{15} cells, which required 53 successive subdivisions ($2^{53} = 2 \times 10^{16}$). Each cell contains 10^{11} atoms and 23×2, that is, 46 chromosomes. Lejeune *et al.* (11) have recently discovered that Mongolism is due to the presence of one supernumerary chromosome (47 instead of 46). Eddington remarked that, on the logarithmic scale, the distances between stars, man, and atoms are equal. If one could "label" all his atoms (by making them radioactive) and could allow them to diffuse freely throughout the world ocean mass, any sample of sea water would contain some hundreds of these original atoms, within the limits of statistical variation.

In 1880, Pfeffer classified living creatures into two large groups: the *autotrophs,* which consume mineral foods only, and the *heterotrophs,* which live parasitically on the former and consequently possess the *minimum* power of synthesis. The plant kingdom consists principally of autotrophs, with the exception of a hundred or so species of bacteria. The number of plants greatly exceeds that of the heterotrophs, mushrooms and animals. Close to four million species are known, of which the insects represent 75%.

The living organism has frequently been approached from the crystal state. If the physical conditions remain favorable, the crystal, inorganic or organic, can grow freely, taking its atoms or constituent ions from the environment. Removed from its mother solution, it retains its structure indefinitely. It perishes only if the physicochemical conditions become unfavorable, by dehydration or fusion for example. A tiny crystal seed may give rise to an infinite number of descendants. The living organism possesses similar properties, as *in vitro* tissue culture experiments have shown. From 1911, A. Carrel cultivated fragments of the heart of a chick embryo, which continued to beat for more than 20 years, even though a chicken only lives for 5 years, and which grew by doubling in volume every 3 days. The volume was therefore increased by 2^{10} in 1 month (10^3) and by 10^{36} in 1 year. Had it been possible to nourish the initial fragment, it would have had, at the end of a year, a mass very much greater than that of the sun.

Bacteria are known that reproduce almost mechanically by dividing in two every 23 minutes like clockwork. In 1897, F. Cohn showed that in a few days a bacteria weighing 10^{-12} g could produce 10^{36} individuals weighing as much as the world ocean. The "explosions of life" which are sometimes seen in the Diatoms testify to this basic characteristic. This explains how emergent life has been able to extend, in an almost explosive manner, to the whole oceanic surface.

Many regions of the globe in which life could not have emerged—deserts, high mountains, inland ice, caverns, and ocean depths—have been colonized subsequently. At a time when no primitive vegetation could have existed in arid regions, the cacti and nontranspiring plant grasses became adapted and established their fauna there.

These considerations should not be lost sight of in discussing the possibility of life on other planets.

The majority of living organisms are indistinguishable within a single species. Distinctiveness comes only after a very long evolution; it is the product of a high degree of complexity and is due to sexual reproduction. It is, however, absent in man in the case of identical twins derived from a single egg. Although these twins are multiple they possess strictly the same hereditary endowment and only between them is tissue grafting possible.

A great many philosophers have been pleased to recognize in the living organism the blossoming of an integral free will, but the facts do not lend support to this belief. The behavior of the majority of living creatures appears to be largely instinctive. In the *Lineola articulata*, *reflexes* are still present in sleeping or decapitated individuals, but *instincts* require psychic integrity and the cooperation of the cerebral ganglia. It is known how human intelligence depends on thyroxine, the hormone secreted by the thyroid gland, and how a great many drugs act on the individual's faculties and behavior. We know that the personality, character, tastes, and capacities of higher living organisms are under the strict control of their internal secretions. The case of true monozygotic twins shows that living creatures are in fact completely determined from the moment of conception. The fertilized ovum contains our whole potential. The chromosomes contain the entire *program* of the individual and may be compared to the magnetic tape of Poulsen's telephonic recorder. Gamow has suggested that this program is written, through the agency of the amino acids, in a binary language, in the same way that the Morse code permits a message in dots and dashes. The naive beliefs of astrology are all the more clumsy in that they insist on bringing the stars' influence to bear at the actual instant of birth, whereas, if these influences existed, they would act on the chromosome arrangement at the moment of conception. Genetic laws show us that these associations are accidental. Parthenogenesis from a single parent is equivalent to the production of true twins. It is analogous to a sexual seedless reproduction in horticulture, by grafting or cutting.

Action of Physical Agents

Of all the physical agents that act on living matter, the most important is heat. Molecular kinetic energy is measured by the product $3/2\ KT$, of Boltzmann's constant ($K = R/N = 1.38 \times 10^{-16}$ erg/degree), by the absolute temperature T. Thus, at the surface temperature of the tropical oceans (27°C), the mean energy of thermal molecular agitation is:

$$W = \frac{3}{2} \times 1.38 \times 10^{-16} \times 300 = 6.2 \times 10^{-14} \text{ ergs}$$

On account of the equipartition of energy, this value is the same for all molecules, the heaviest being the slowest. Converted to electron volts, this energy is equal to $6.2 \times 10^{-14}/1.6 \times 10^{-12} = 0.04$ ev. This is a temperature close to the vital optimum.

The vital activity has, for a long time, been calculated from Van't Hoff's empirical law governing the rate of a chemical reaction as a function of temperature. This law states that the rate doubles for each 10°C rise in temperature, but it is valid only within narrow limits. Langevin and Rey (12) arrived at it again from kinetic theory. The distribution of molecular velocities obeys a law of Maxwell and the ratio of the rates of dissociations brought about by impacts at temperatures T_0 and T is:

$$\nu_T/\nu_{T_0} = \sqrt{\frac{T}{T_0}} \exp\left[A\left(\frac{1}{T_0} - \frac{1}{T}\right)\right]$$

If T is close to T_0, Van't Hoff's law is again established:

$$\nu_T/\nu_{T_0} = \exp\left[A\left(\frac{1}{T_0} - \frac{1}{T}\right)\right]$$

Thus, between $t_0 = 20°\text{C}$ and $t = 20°\text{C} + 10°\text{C}$, that is: $T_0 = 293°\text{K}$ and $T = 303°\text{K}$, we get: $\nu_{303}/\nu_{293} = 2$. From this, $A = 6000$. It follows from this that the vital activity of the bird ($t = 42°\text{C}$) is 1.36 times (greater than) that of mammals ($t = 37°\text{C}$).

The rate is 2^6 times, that is 64 times, greater at 60°C than at 0°C. A reaction which would take one day at 30°C would be complete in 4½ minutes at 150°C, in 4 months at —20°C, and in 10^4 years at —100°C. To all intents and purposes, cold prevents chemical reactions. This is the same principle as Tellier's on the suspension of fermentation by refrigeration. The mammoth of Berezovka (1901), preserved in the Siberian permafrost, was still edible after 12,000 years. If the vital activity is put into suspension in cold-blooded organisms, the vital organization is not destroyed. Many fishes and batrachians can be frozen without being harmed. This is *anabiosis*. Snails have been frozen to —100°C and brought back to life. Bacteria and many germs, cysts, spores, pollen grains, eggs, and seeds persist for thousands of years, neither alive nor dead (the Ptolemaic papyrus of Gallipa, 1919), and, as the experiments of Becquerel (13) have shown, withstand complete desiccation in a high vacuum and when kept for a long time at the temperature of liquid helium (+0.7°K). Colloidal gels likewise remain unaltered under these conditions.

However, although latency is reversible, death is irreversible.

On the other hand, although some crystal lattices, such as that of graphite, withstand the temperature of incandescence, a slight rise in temperature is fatal to all living organisms, as Appert showed in 1810 when he introduced sterilization, which occurs at about 120°C. At this temperature, the molecular collisions had sufficient energy to destroy the fragile macromolecular structures. The mean energy equals:

$$W = \frac{3}{2} KT = \frac{3}{2} \times 1.38 \times 10^{-16} \times 393 = 8.1 \times 10^{-14} \text{ ergs}$$

Thus, in the range of temperatures prevalent in the universe, from 20 million degrees at the center of the dwarf stars to the neighborhood of absolute zero at the surface of the distant planets, life can continue to exist only in a tiny region that is even much more restricted than that in which organic chemistry operates.

All the other causes of mechanical destruction, crushing, hyperpressures and ultrasounds, have the same effect. Ruska has shown that ultrasounds break the tobacco mosaic molecule into fragments that lose their virulence. Intense gravitational fields do not seem to be fatal to all living organisms, but it is clear that the higher creatures would not be able to withstand them. For physiological reasons the higher creatures are unable to tolerate strong prolonged accelerations and, perhaps, prolonged weightlessness.

Light becomes lethal for living organisms when its $h\nu$ quantum is greater than 4 ev, that is to say, when the wavelength (3200 Å) belongs to the ultraviolet range. Remember that electromagnetic radiations have no *direct* action on matter and that their physical and biological effects (vision, phototropism, erythema, etc.) are due solely to the photoelectrons which they liberate when absorbed.

In water, photoelectrons of a few volts travel only a very short distance, of the order of a hundredth of a micron. All their energy, that is, 6×10^{-12} ergs, is liberated locally in the form of a "hot point" equivalent to an extra-strong molecular impact. The abiotic action of ultraviolet light shows a spectral distribution corresponding to the ultraviolet absorption spectrum of thymonucleic acid, the sensitive element of the nucleus. Figure 1-11 represents, after Coblentz (1934), the erythemal spectral curve E and the sterilizing curve S. Spectroscopists are well aware that gelatin has its absorption maximum at about 2650 Å and that it became quite opaque at 2350 Å. In order to work in the far ultraviolet, it is necessary to use emulsions containing very small amounts of gelatin (Schuman plates). A 200-roentgen dose of X- or γ-rays is immediately fatal.

High quantum corpuscular rays, that is, β-rays, those of Lenard, protons, α-rays, the mesons of cosmic radiation, etc., all have a similar action but, instead of uniformly destroying all living matter, as heat does,

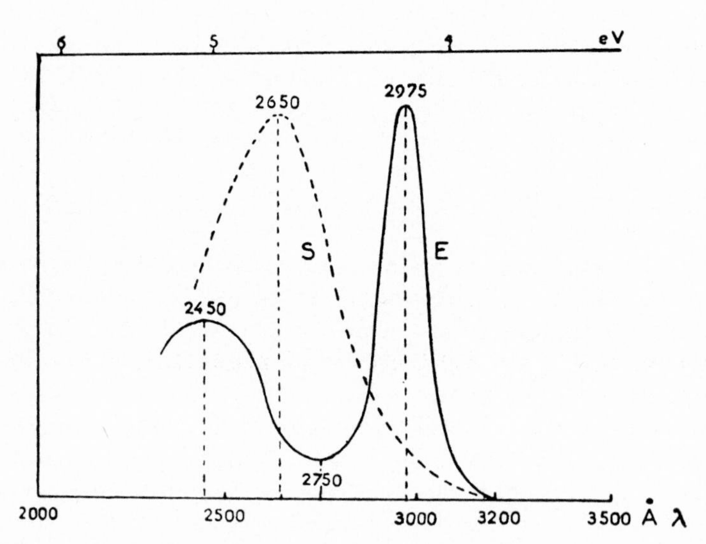

Fig. I-11. Erythemal spectral curve E and abiotic spectral curve S in the ultraviolet (after Coblentz).

they are able, because of their rare spatial distribution, to produce specific intracellular effects like those revealed in F. Holweck and A. Lacassagne's experiments: suppression of motility and reproduction, and postponed or immediate death. We shall return to these effects.

The electron bombardment of organic matter, as, for example, in the electron microscope, leaves a fixed carbon skeleton, as we have shown spectroscopically (1926), in obtaining pure K α radiation (45 Å) from this element, starting with any material target in the presence of one or other carbonaceous vapor.

Although organisms are transparent to γ- and X-rays, they are opaque to neutrons because of their richness in hydrogen: A radiograph taken with neutrons would give the inverse image of a radiograph taken with X-rays. The organisms absorb large numbers of neutrons and this is the cause of the artificial radioactivity of some of their chemical elements —sodium, silicon, and phosphorus. A 10^5-kw nuclear reactor can kill a man in 2 minutes within a 100-meter radius through its neutron flux, and the tolerance dose, for thermal neutrons, is estimated at 2000 per square centimeter of irradiated surface per second.

In recent years, it has been recognized that mammals, such as bats, use ultrasounds for navigation, and that a number of arthropods—insects, crustaceans, and arachnids—utilize natural polarized light to orient themselves. Images, colors, and sounds, it is known, do not exist in nature. Living organisms manufacture them for their own use.

Electric and magnetic fields seem to have no action on living matter. The natural field of the troposphere, in the neighborhood of the ground, is very variable and is of the order of 100 volts per meter. The geomagnetic field is close to ½ gauss. No organs sensitive to these fields are known. However, Rocard (14), experimenting with *water-diviners,* observed that some subjects were sensitive to single, weak, magnetic fields (0.3 milligauss).

The sense organs are instruments of a sensitivity unequalled by human industry that could not be surpassed. A 2-photon flash of 5000 Å can be perceived by the human eye. A single odorous molecule may be detected by an animal with a developed olfactory sense. We may mention the odorous marking of ants' tracks. Much animal behavior, popularly—and too readily—qualified by the generic term "instinct," must be associated with *olfaction* and with the reflexes which it triggers off. Nutrition, reproduction, and many tropisms are controlled by olfaction and so their mechanism appears to be purely chemical.

Naturalists have seemed surprised that higher animals do not recognize themselves in a mirror: This is because their method of identification is not by sight but by smell.

This subtlety of the senses, which functions on the molecular level, will mean that the robots of cybernetics, however perfectly they may be imagined, will never, because of the *scale* of our physical instruments, be more than enormous, crude machines.

References

1. H. Le Chatelier, "Leçons sur le Carbone." Hermann, Paris (1908).
2. R. Audubert, "Des précipités colloidaux aux macro-molécules." Presses Universitaires de France, Paris (1956).
3. F. Sanger and L. F. Smith, *Endeavour* **16,** 48 (1957).
4. J. B. S. Haldane, *New Biol.* **16,** 12 (1954).
5. J. D. Watson and F. H. C. Crick, *Nature* **171,** 737 (1953).
6. P. Douzon, J. C. Francq, J. Polonsky, and C. Sadron, *Compt. Rend.* **251,** 976 (1960).
7. H. T. Nixon and R. D. Woods, *Virology* **10,** 157 (1959).
8. J. Bronfenbrenner and P. Reichert, *Proc. Soc. Exptl. Biol. Med.* **24,** 176 (1936).
9. H. L. Fraenkel-Conrat and R. Williams, *Proc. Natl. Acad. Sci. U.S.* **41,** 690 (1955).
10. W. T. Astbury and C. Weibull, *Nature* **163,** 280 (1949).
11. J. Lejeune, M. Gautier, and R. Turpin, *Compt. Rend.* **248,** 602 (1959).
12. P. Langevin and J. Rey, *Radium* **10,** 142 (1913); C. E. Guye, *Arch. Sci. Phys. Nat.* (*Geneva*), **5,** 17 (1935).
13. P. Becquerel, *Compt. Rend.* **190,** 134 (1930).
14. Y. Rocard, "Le signal du sourcier." Dunod, Paris (1962).

II
THEORIES OF THE ORIGIN OF LIFE

Before discussing the problem of the origin of life, it is important to be certain that the problem exists. Many learned men, among whom may be cited Spinoza, Buffon, Lord Kelvin, Liebig, Pasteur, and Arrhenius, have thought, for various reasons, that this was a false problem, that life was as eternal as the world itself, and that, therefore, it had never had a beginning. This was Pasteur's opinion when in 1878 he wrote:

"Spontaneous generation—I have been seeking it without finding it for twenty years. No, I do not think it impossible. But whatever justifies us in insisting that it has been the origin of life? You place matter before life and you cause matter to exist through all eternity. Who can say that the ceaseless advance of science will not oblige the learned men who will be living a century, a thousand years, ten thousand years hence . . . to affirm that life, and not matter, has existed through all eternity?"

Pasteur, following Laplace, undoubtedly believed in the immutable world, in the sun as an inexhaustible and indefinite source of heat and light, and in the unchanging solar system. Thus, the problem of the origin of life at once assumes a cosmic character. There assuredly exist many phenomena outside sensitive matter as electromagnetism and the propagation of light according to Maxwell's theory testify. But a "life" of this nature, assuming that it exists, would be so far from what we know that it would be necessary to give it another name. An electromagnetic type of psychism, without material support, would be from within the realm of fiction. It would require a cosmic source of energy and could not be characteristic of an individual.

This mystical conception of life has always dominated philosophical thought. Life was placed on another plane to matter. Man, discouraged by its complexity, could only fall back on the supernatural. What was more obvious than to attribute to each living creature, to each plant, a tiny "genie," intangible and intelligent, controlling its growth, evolution, and reproduction. Indeed, finalism seems to appear everywhere in the living world and an enormous scientific effort must be made in order to get rid of it. Pagan animism placed a spirit not only in each living being but in the crags, lakes, forests, and the whole of nature. This poetic conception of a world of fairies is still deeply rooted. Poets certainly cannot do without it, but men of science are more exacting. They are forbidden to play with words and to multiply creatures unnecessarily.

In 1790, Kant had a supernatural and finalist conception of life. Nowadays, not only philosophers such as A. Comte, Bergson, and Le Roy, but paleontologists, such as Teilhard de Chardin, and biologists, such as Vandel and H. Rouvière, look upon it with favor. Teilhard de Chardin wrote in 1931, in "L'Esprit de la Terre":

"The Cosmos . . . is fundamentally and in the first place living and its whole history is, at bottom, only an immense psychic affair: the slow but progressive assembling of a diffuse conscience. . . . The young Earth contained a quantum of conscience and this quantum has passed in its entirety, into the Biosphere."

A. Vandel, in "Le Message de la Biologie," expressed a similar thought in 1943:

"This hypothesis implies that matter would not have been, at the beginning, what it is today. It was a matter rich in power and unexpressed possibility, capable of giving rise to both the organic and the inorganic, the living and the inert. In giving birth to life, it would have caused the essential part of its creative energy to pass into it. This effect would have reduced matter to a degraded, devitalised residue which had lost the majority of its ancestral qualities.

"It is this matter which the physicists and chemists are studying today. It would be as vain to wish to draw life from this matter, exhausted and emptied of its primordial energy, as it would be to hope to extract, from bran and faints, the seed of corn, the harbinger of future harvests."

Occultism is still, nowadays, even among learned men and eminent specialists, as popular as in the Middle Ages, and one should not be astonished at this, for the supernatural, the incredible, spiritualism, spiritism, and vitalism will always have adherents. A number of philosophical doctrines allow of sentient "lives" without material support: the life of spirits, of deities, of angels, and of demons. The "spirit" is incarnated, as electricity charges the condenser. The most delirious lucubrations are permissible in this domain of metaphysics. Meanwhile mystic interpretations only conceal our ignorance and do not advance our knowledge one step toward a rational solution. They are a defeat of intelligence.

The *order* and *scale* of the phenomena should also be respected. Thus J. Rostand (1) declared:

"It might also be thought that the genesis of life could have occurred only during a specific period of cosmic evolution and, for example, at a certain phase of the expansion of the Universe."

Now, it is known that this expansion is not revealed either in the metagalaxy or, *a fortiori,* in our galaxy, where the mean density of matter is greater than Einstein's critical density.

The problem of the origin of life is also a false problem for the supporters of cosmic panspermia, such as Lord Kelvin. Arrhenius declared: "Matter, energy and life have merely changed form and place in space."

Van Tieghem, in his "Traité de Botanique," wrote, in 1891: "The terrestrial vegetation has had a beginning and will have an end but the vegetation of the Universe is as eternal as the Universe itself."

They suggested that life was only propagated from one celestial body to another.

A third attitude again denied the admissibility of the problem of the origin of life. A number of learned people are disconcerted at the search for first causes. In 1929, Vernadsky declared in his "Biosphère" that research on the cosmic era of the earth and the genesis of life are useless, illusory, and indeed, harmful and dangerous. This has been said before, in every epoch, about questions that scientific investigation has still not elucidated. It is a sterile attitude, the prudence of which may appear excessive, and such reserve only hides a lack of enthusiasm, curiosity, and imagination. For E. Renan (1845), on the other hand, the passage from the "gross to the vital" was the essential problem of science. But in 1863, Darwin was still judging it to be" unreasonable to think about the origin of life." The attempt to assemble the known facts into a coherent theory of the origin of life is related to the program already outlined by Lucretius in his fine poem "De Natura Rerum," and its meaning, from the viewpoint of general philosophy, is clearly apparent. As R. Virchow said in 1865: "If life has had a beginning, it should be possible for science to determine rationally the conditions of this beginning."

To understand life in the universe, we must, in this field as in so many others, cast off our geocentric and anthropomorphic mentality and consider the problem on the cosmic time scale, where the unit is a thousand million years. This unit, which is very long for us, since it has allowed the living terrestrial world to emerge and evolve, is actually very short on the scale of the universe. It is only of the order of the age of the actual radioelements and stars. On a planet, several thousand million years probably represent the average duration of life permitted by the frequency of cosmic catastrophes (novas, collisions, etc.) and the natural evolution of the galaxies. If it is shown, by plausible cosmogonic interpretations, that, contrary to Laplace's opinion, the solar system was formed four thousand million years ago, the problem of the genesis of life on our planet clearly arises. But first, it is instructive to outline, in brief, the history of the rational hypotheses that have been put forward to resolve the problem.

Spontaneous Generation

Among the ancients, who sometimes were not very exacting observers, the doctrine of *spontaneous generation,* that is, without parents, of small living creatures from Aristotle's four elements—earth, water, air, and fire—was universally accepted. Lucretius himself saw numerous animals emerging from the earth "engendered by the rains and the heat of the Sun."

But the size of these living creatures has constantly diminished as observation improved. Remember the controversy between Van Helmont and F. Redi in the sixteenth century. After the discovery of microscopic life by Leeuwenhoeck in 1663, they were moved back into the infravisible region until the experiments of Schwann, and then of Pasteur, showed the emptiness of this belief. In the eighteenth century, Needham (1760) and Buffon were in controversy with Spallanzani; later Pouchet (1859) took issue with Pasteur, and Charlton Bastian (1872) disagreed with T. H. Huxley. It was this problem of spontaneous generation which led Pasteur to discover Bacteriology.

In our time, this hypothesis was carried by B. Moore, R. Esnault-Pelterie and Prenant into the submicroscopic region, to the virus scale, until the advance of electron microscopy showed, once again, the correctness of Harvey's adage (1619): "*Omne vivum ex vivo.*"

Esnault-Pelterie (2), who considered that there was no absolute division between organic matter and organized living matter, emphasized the degree of complexity needed to ensure its organization and calculated the lower limit of the mass of a living organism. It is in this frontier region between the macromolecules and the smallest living creatures, in the domain of the phages, that the origin of life must be sought. The latter must thereafter be born continually: "I am convinced," he said, "that it is there that the origins of life must be sought. I am convinced that it is *there* that it came into being by equally imperceptible degrees. I am convinced that life, by sequences of pure physicochemical phenomena, continues to be born every day from the inert, under our blind eyes and our feeble microscopes, and I will give this theory the name of physicochemical *aidiogenesis* or perpetual creation."

This idea had already been expounded by B. Moore. The erudite engineer adds (2): "The probability that a phenomenon as widespread as life has an adventitious, that is to say, improbable, origin is certainly very low. Accordingly I grant, with little chance of error, that this origin is a phenomenon as little exceptional as life itself."

We know however that the viruses and phages are inert *in vitro* and multiply only at the expense of a living host. We also know that these asymmetric macromolecules could only have originated as a result of improbable dissymmetric physicochemical influences which are no longer encountered in nature: the two enantiomorphs would be equally abundant, like right and left quartz crystals. Finally, all the nonliving organic matter that we know of, such as resins and urea, are waste products of life. In present-day nature, as R. Dubois noted as far back as 1914 in "La Vie et la Lumière" ". . . organic compounds, such as tartaric acid, simpler than living organisms, are not seen to form spontaneously."

The reason is that these compounds are composed of endothermal substances which require, for their synthesis, an external energy, such as a high temperature, electrical discharges, radioactivity, or high-frequency light. The traces of organic substances resulting from volcanic chemistry, such as the sulfocyanates, produced by the reaction of NH_3 with CO_2 and H_2S, are not stable and never accumulate. Of course, rare iron and calcium oxalates and calcium and aluminum mellates are known, but they probably result from the action of the mellitic acid of lignites on limestone and alumina. It is unlikely that mellitic acid results from the electrolysis of water by telluric currents between electrodes of natural graphite as may be accomplished in the laboratory.

Henriet (3) has reported some traces of formaldehyde in the atmospheric air. According to Dahr and Ram (4), rainwater may contain from 0.1 to 1.2 mg per liter. The formaldehyde seems to be of biogenic origin rather than arising from a controlled oxidation of methane by ozone. Ozone completely oxidizes the hydrogen, methane, and ammonia of volcanic or biogenic origin. Lightning only generates ozone, oxides of nitrogen, and ammonium nitrite, but no organic substances. All the compounds known to us, coal, lignites, oil, bitumens, waxes, and resins, are waste products of life. Any trace of young, sterile, assimilable organic matter would immediately be the prey of the bacteria that are abundant everywhere, as Pasteur has shown.

That then is one doctrine: spontaneous generation, which is being withdrawn step by step as observation improves. Observation has now reached the frontier of the living world, the last area below which the living creature gives way to the macromolecule. What hereditary patrimony could creatures, born spontaneously, without parents, possess? The mirific results that have been announced with such enthusiasm in various countries from time to time in an effort to refute Pasteur's work, do not deserve—and we shall see why—to be believed at all.

Cosmic Panspermia

In the face of the insurmountable difficulty which the problem of the origin of life seems to present, many authors have suggested that life came to us from other celestial bodies, and such well-informed men as H. von Helmholtz, Lord Kelvin, and Arrhenius have subscribed to this opinion. In 1821, de Montlivault attributed the seeding of the earth to the carriage of germs of lunar origin which had been ejected by volcanoes. However, volcanic exhalation can only be sterile, and for a very good reason. Kelvin, who felt that the collisions between celestial bodies played a leading role, wrote seriously in 1871: "When two worlds meet each other at a velocity of 30 km/sec the impact is so violent that an innumerable quantity of fragments are hurled into space. These fragments must bring germs onto the planets on which they land."

With such velocities, it is clear that the transformation of the kinetic energy into radiation leads, at very high temperatures, to a complete volatilization with formation of gas and cosmic smoke.

In 1865, Richter (5) assumed that germs were carried by meteorites; the so-called "carbonaceous meteorites" are of special interest. They are extremely rare and only eight are known to have fallen. They are of a low density, between 2.9 and 3.0 g/cm.^{3} The stone of Orgueil, which fell on the fourteenth of May 1864, is a peridot containing iron and nickel sulfides, iron and magnesium carbonates, ammonium chlorhydrate, and 15% of hydrated organic substances. On analysis, Cloez (6) found that the latter had the composition: C = 63.45, O = 30.57, and H = 5.98%, which is similar to that of humic substances, ozocerite and lignites. But that does not mean that these meteorites have brought residues of biogenic origin, for these phosphorus and nitrogen-free substances are analogous to volcanic bituminous exudates. The obvious thing to do was to study the carbonaceous parts bacteriologically and microscopically, although it was impossible to collect meteorites aseptically. It was also necessary to verify the sterility of the inner fissured regions. Meunier referred fresh specimens from the Orgueil meteorite to Pasteur, but the aseptic samples gave only negative results which were not published. On the other hand, Smith, Berthelot, Schutzenberger, and Bourgeois showed that the hydrocarbons of the carbonaceous meteorites were analogous to those of white iron. They are produced by the reaction of water vapor on iron carbides at high temperature.

The Cold Bokkeveld (Transvaal) meteorite contained a bituminous cement; extracts of which did not show optical activity characteristic of a biogenic origin (Müller, 7). Calvin (8) reported the presence of heterocyclic compounds, pyrimidines, and purines, within a stony meteorite. We shall see, in discussing their origin (and the presence of the NH and CN radicals in cometary spectra), how the paleovolcanic chemical evolution of the lithospheres of terrestrial planets led—in the absence of free oxygen—to pyrogenous syntheses of such compounds which have permitted life on earth and which have remained occluded in the lithospheres of Mercury, Mars, the moon and Olber's planet, from which the cosmic rocks originate.

Periodically, uninformed authors announce loudly, that they have discovered vestiges of biogenic origin in meteorites. Nagy *et al.* (9) stated that they discovered biogenic substances such as cholesterol in the Orgueil meteorite. According to Astre (10), Douste-Blazy confirmed at Toulouse, the absence of lipids. Claus and Nagy (11) have also claimed the discovery in the Orgueil and Ivuna (1938) meteorites, of nine new kinds of indigenous microfossils of dinoflagellates for which they have proposed a new system. Deflandre (12), examining the preparations and photomicrographs of Claus, found no biogenic vestiges and concluded severely: "The inexperience of those responsible for the descriptions is manifest in certain diagnoses."

Meteorites have, moreover, a characteristic mineralogical structure. They contain minerals only: metal carbides, graphite, diamond, sulfides, phosphorides, chromite, magnetite, anorthite feldspar, pyrite, hornblende, amphibole, serpentine, labradorite, analogs of the deep terrestial rocks of igneous origin, and do not contain any rock that has issued from sedimentary terrains. This suggests that they came from a celestial body (the planet postulated by Olbers between Mars and Jupiter), which has never possessed oceans or atmosphere, and was broken up at the lunar stage, without, consequently, ever possessing a biosphere. Meteorites all belong to the solar system, as their age and isotopic analysis of their elements show, and none has come to us from another planetary system. In fact, the passage through the atmosphere of a meteor moving at velocities varying between 12 and 72 km/sec is accompanied by a considerable rise of surface temperature and by a strong deceleration that would cause it to break up.

Arrhenius (13) took up the hypothesis of cosmic panspermia again in another form. According to him, cosmic germs from other planetary systems were expelled by the pressure of the stellar radiation and reached the earth on interstellar cosmic dust. This hypothesis can only be maintained with the aid of a succession of excessive "shoves." Let us consider, for example, the departure from the earth of a naked living germ produced by the biosphere. First of all, it must escape from the atmosphere. Now, neither Brownian movement nor ascending currents are capable of accomplishing this. To overcome gravity, the author attributed a negative electric charge to the germ and assumed that it was subject to the terrestrial electric field. But this terrestrial electric field is restricted to the troposphere and it is reduced to zero well before the conducting ionosphere is reached. In order to expel the germ from the solar system, the author then utilized the solar radiation pressure. For a granule that has the mean density of the earth, 5.5 g/cm^3, gravity and solar radiation pressure are in equilibrium when the diameter attains 0.3 μ. At the density of water, it will be expelled by the radiation pressure. This type of thinking was fashionable at this period. The shape of the sun's corona and the repulsion of cometary tails by the sun were attributed to it. In fact, it is an extraordinarily weak force, no more than 4×10^{-5} dyne/cm^2 on earth. We have shown (14) that the repulsion of the cometary tails was due not to this force, but to an incomparably greater reaction force caused by the escape of thermal gas and the evaporation of the cometary dust clouds.

For a germ of foreign origin to land and seed the earth, gravity must exceed the solar radiation pressure. With this in view, the author allows that it might unite with a granule of cosmic dust. However, the dust that exists in the solar system or in interstellar space would certainly not be a vehicle for the germ but rather a cause of destruction, since the high velocities of the particles confer a high kinetic energy on them. A germ entering a planetary atmosphere at high speed in this way would, moreover, be burnt up immediately like a shooting star.

Esnault-Pelterie (2) has subjected Arrhenius' hypothesis to calculation. This scientist considered an absorbent germ whose *terrestrial weight* would have been equal to the *pressure* of solar light at the level of the terrestrial orbit. It would be expelled from the solar system with an increasing velocity. The acceleration, at first zero, passes through a maximum and falls away again to nothing, that is to say, the final velocity tends toward a limiting value. If x is the distance to the sun, in astronomical units then:

$$\nu = 1719 \sqrt{1 - (1/x)} \text{km/sec}$$

Initially, $x = 1/3$ and $\nu = 0$. At the distance of Mars, $x = 1.52$ and $\nu = 10^3$ km/sec, and at infinity, $\nu = 1719$ km/sec. The germ has become a cosmic projectile, but if its mass is calculated, it is seen that its size would be that of a molecule (5 Å) and not of a germ. A 0.5-μ granule would have a terrestrial weight much greater than the radiation pressure and would have to have been a considerable distance from the earth initially. Its limiting velocity would then attain 35 km/sec. In actual fact, such a germ would be transparent, and diffraction would reduce the thrust even more. Esnault-Pelterie showed that the trajectory of the particle must fulfill conditions which are highly unlikely to pass from one planetary system to another, given the extreme stellar rarefaction in our region of the galaxy. The orbit must be exactly oriented and the probability of this is no higher than 10^{-15}. The stellar movements themselves must also be accounted for. The recipient star must possess at least one planet on which the physical conditions are compatible with life. Finally, the trajectory must lie within the orbital plane of the planet and the planet must coincide with the germ.

In fact, Pasteur's experiments showed that the density of the atmospheric germs decreased extremely rapidly with altitude. During the ascent of the American manned balloon "Explorer II" to an altitude of 22 km in 1935, only 10 microorganisms, belonging to various species, were collected in 66 m^3 of air, although 15,000 bacteria per cubic meter of air and 130,000 per gram of dust were found in Paris. Germs rising to the atmospheric ozone layer at about 25 km would be destroyed there by this gas and, at a higher altitude, by the abiotic solar ultraviolet and cosmic radiation. Becquerel (15) has carried out numerous experiments on this subject with the most highly resistant dried spores, *Aspergillus, Penicillium, Mucor, Mugatere,* etc. He showed that, if they withstood a high vacuum in the region of absolute zero, then 6 hours exposure to the radiation from a quartz mercury vapor lamp was fatal. Henri and Moycho (16) have shown that the bactericidal effect, at a maximum when $\lambda = 1800$ Å, occurs with a dose of only 2×10^4 ergs/cm^2. Now, spectroscopic measurements carried out in this region by means of rockets have revealed an intense chromospheric spectrum.

The theory of cosmic panspermia involves the *uniformity* and the *universality* of the forms of life and living species throughout the universe. Everywhere, molecular dissymmetry would be represented by the same enantiomorphs. The theory is not without grandeur, but it remains disconcertingly weak. It is a mere dismissal which seeks to avoid the true problem of the origin of life. If, despite everything, one insisted on preserving it, it would be necessary to invoke the transport, spontaneous or otherwise, of germs by astronauts. We shall see that there are innumerable planets peopled with intelligent beings and it is certain that a crowd of spaceships has been wandering in the galaxy for thousands of millions of years, carrying their pilots' remains. But these spaceships collide with celestial bodies and disappear into cosmic smoke and a safe landing for them seems to be out of the question. Even if the landing were safe, the germ abandoned on the primitive earth would have to find the best suitable state for its reception, a state which we shall be studying. The fact is that such a hypothesis is highly improbable, and we shall have no need of it. Paleontological evolution shows that life was not brought directly to the earth from another celestial body in an already evolved form, but that it began here with very simple primitive forms which continued to become more complex.

Creative Chance

According to the ancient thesis of the "exhaustion of possible fortuitous combinations," for which Empedocles (490–444 B.C.) and Herodotus (484–406 B.C.) seem to be responsible, chance is capable of effecting the most improbable combinations, given sufficient time. Herodotus declared: "Length of time may bring anything to pass."

This thesis was revived by Huxley and E. Borel with their "typing monkeys," who struck haphazardly for a very long while and finally reproduced Pythagoras' theorem.

This thesis has the advantage of lending itself to probability calculations. So it is that there is a chance that n white balls and n black balls, agitated and mixed in a die, will be separated, on playing one game a second, at the end of a time $\theta = 2n!/n!n!$. If $n = 10$, this time will be about 2 days. If $n = 20$, it will be some 3000 years, and if $n = 100$, it would be necessary to wait 10^{51} years. This thesis has been discussed at length by Guye (17). In two theoretical works of great interest, the famous physicist remarked that it is the *scale* of observation which constitutes the phenomenon.

"Let us consider," he said, "a uniform mixture of H_2O, CO_2 and N_2 gases and of some other constituents of living matter. Let us suppose that this mixture, previously sterilized by a raised temperature, is then subjected—for example, through a quartz window allowing the transmission of ultra-violet rays—to the action of sunlight—not for a small number of days or years, but for an immense number of centuries. We shall then be able to grant that all the dissociations, combinations and associations of atoms, molecules and electrons, which are capable of being produced under the influence of light and thermal agitation, which are, in a word, *possible*, will finally occur in the course of the ages.

"At one point of the volume, an association corresponding to the physicochemical constitution of living matter will therefore be able to be produced.

"In this hypothesis, the emergence of life would be due to a *fluctuation of a very rare type*, that is to say, to what we have familiarly called an extraordinary chance. Note however that if we are permitted to make this assumption, it is because we know nothing, so to speak, of this passage from non-living to living matter. The only more certain thing that we know is that, on the one hand, we have never been able to produce it in our *vitro* reactions and that, on the other, we have never been the witnesses and spectators of such a transformation, two conditions which are in practice compatible with the hypothesis of a very rare fluctuation."

The famous physicist appealed to two distinct causes: the movement of thermal molecular agitation, which is unable to bring about any chemical reaction in this stable gaseous mixture, and the photochemical reactions caused by ultraviolet light. But, as we shall see, Berthelot and Gaudechon showed in 1910 that the solar light had no effect on such a mixture. Nevertheless, we can, more correctly, envisage that of an ultraviolet radiation in the region of 2000 Å in order to bring about synthesis while preventing photolysis. But many other conditions will have to be satisfied to arrive at compounds as complex as the nucleoproteins, and it is only on the scale of the *laboratory of nature* that such a phenomenon can be envisaged. Besides, in our problem, time is not limitless. We have only a few thousand million years available, and these probabilistic considerations fade away completely in the face of such a short period of time.

Guye pursued his subtle analysis by referring to Pasteur's work on the molecular dissymmetry of living matter and of P. Curie's (1894) work on the role of dissymmetry in physical phenomena. He concluded that, if the physicochemical evolution, governed by Carnot's principle, is due to statistical effects, then the cause and the origin of life and thought must be sought in individual molecular effects.

"Heterogeneity," he said, "is the creator of dissymmetry and dissymmetry is the creator of force." Applying the laws of combination analysis (Stirling's formula) to simple cases of the production of configurations in heterogeneous powders formed of two or three homogeneous constituents, he showed that random mechanical agitation was powerless to cause an increased dissymmetry to appear when the number of grains was large. It is necessary to appeal to other causes of organization, such as the action of chemical valencies and photochemical effects. Organized matter is made up of macromolecules having dissymmetric structures which alone permit the appearance, on our macroscopic scale, of the individual molecular properties that are characteristic of life.

If, by any remote chance, in the experiment proposed by Guye, protein macromolecules were formed after many millennia, they would not have the characteristic asymmetry of genes. They would not have *become organized* and would not have become living matter, and, even if they had, they would have been destroyed immediately by the abiotic radiation utilized.

It is this that Lecomte du Noüy (18) failed to understand when he wished to show, in reproducing Guye's calculation, that life could not have appeared of itself but that it had to have been created by a supernatural cause. Dissymmetry may be expressed by a number between ½, which is homogeneity, and 1, complete heterogeneity. A merely binary molecule, made up of 20,000 atoms, having a molecular mass of the order of 10^5 and endowed with a symmetry of 0.9, would have a probability of spontaneous appearance of 2×10^{-321}, or, in other words, absolutely none as far as we are concerned. But atoms and molecules are not the simple black and white balls of the theory. If one wished to construct the simplest of cubic lattices, that of rocksalt, with such balls representing the Na^+ and Cl^- ions, a thousand years agitation would never achieve it. But if the two types of ions meet in a saturated solution, the crystal seed will straightaway appear.

Nonetheless, certain crystallizations are very improbable, such as the famous crystallization of glycerine that appeared during the winter of 1867, or the crystallizations that occurred spontaneously at the center of a limpid glass block, melted at Nancy in the last century (E. Desguin).

A great many authors have been even more ambitious than Guye, and have not been afraid to suggest that life appeared directly in an already highly organized form. Osborn (19) made life appear with S. Winogradsky's autotrophic bacteria. For him, life was born not in the "oceans, still nitrogen-poor and with little salt" but in the continental fresh waters. The primitive atmosphere contained free oxygen and the carbon dioxide had been reduced by "other than chlorophyllic agents and by purely chemical reactions." Constantin (20) and Perrier saw the emergence of the living world in the spontaneous, almost miraculous, appearance of chlorophyll. As B. Moore remarked, this substance, of a hitherto unparalleled complexity, still required the previous existence of a highly evolved living substrate.

Bacteria that are autotrophic although they do not contain chlorophyll are much more complex than the heterotrophic bacteria and consequently seem to be of later evolution. Their energy deficit is corrected by an exothermic reaction which is still an oxidation. We may cite:

1. The oxidation of NH_3 into nitrites by *Nitrosomonas.*
2. The oxidation of nitrites into nitrates by the *Nitrobacter.*
3. The oxidation of ferrous oxides into ferric hydrate by *Spirophyllum ferrugineum. Chlamydothrix* (*Leptothrix*) *ochracea* is only facultatively

autotrophic and always needs traces of organic substances for its nutrition.

4. The oxidation of H_2S by the sulfobacteria, the most widespread of which is *Beggiatoa alba.* The sulfur produced appears in the form of granules and may be oxidized in its turn.

5. The oxidation of hydrogen: *Bacillus pantotrophus* and *Bacillus picnoticus.*

6. The oxidation of methane: *Bacillus methanicus* (Söhngen), which can be cultivated in an environment not containing organic substances, with an atmosphere of CH_4. A *Methanomonas methanica* is also known. *Bacterium hexacarbovorum* utilizes not only CH_4, but also toluene, xylene, etc. Some common bacteria are able to oxidize methane but are only facultatively autotrophic. Champagnat has proposed the use of bacteria, such as *Pseudomonas,* to remove paraffin hydrocarbons from oils in order to lower their freezing point while supplying nutritional proteins.

Apart from these cases, the only autotrophs are organisms with chlorophyll or an equivalent pigment such as bacteriopurpurine. Bacteria that utilize hydrogen, methane, and ammonia do not even take it from vulcanism but from the biosphere. In the absence of the biosphere, bacteria would have to be able to live at the expense of volcanic exhalation, but this never produces free oxygen, and free oxygen did not exist in the primitive atmosphere. They would also have to decompose CO_2 or H_2O in order to respire. In fact, any autotrophic organism transported to a primitive sterile earth, or to Mars or Venus, would perish because of the failure to find respirable oxygen. In the end, only the ferrobacteria would survive.

However, it cannot be accepted that bacteria do not respire, that is to say, that they do not oxidize any of their organic matter. To conceive of the formation of ferrobacteria *de novo,* it would be necessary to have the simultaneous occurrence of: (*a*) the oxidation of ferrous oxides into ferric, (*b*) the reduction of atmospheric CO_2, and (*c*) the synthesis of organic substances.

In the absence, and even in the presence, of free atmospheric oxygen, such an energy balance is in deficit. The appearance of life is inconceivable in the absence of photochemical dissociation of carbon dioxide or of water.

The autotrophic bacteria, far from being primitive, seem rather to be parasites using the waste products of life. It is superfluous to mention the hypotheses about the almost miraculous appearance of much more highly evolved organisms such as the Cyanophyceae and single-celled Algae.

The Synthesis of Organic Matter

If life has not appeared by chance, it is because of the normal evolution of energy processes which have created, at a certain epoch of the earth's history, an abundance of organic matter, which was at first sterile.

In 1875, Pflüger (21) based a theory of the origin of organic matter on the synthesis of cyano compounds, which would have occurred during the period of incandescence of the globe.

Gautier (22) thought he would find, in the high temperature of volcanic chambers, the site of these pyrogenic syntheses. It is known that formic acid can be produced by Berthelot's reaction:

$$CO_2 + H_2 \rightleftharpoons H{\cdot}CO_2H$$

A mixture of carbon monoxide and hydrogen gives, at 400°C, C, CO_2, CH_4, H_2O, and some traces of formaldehyde. The action of H_2S and CO_2 on NH_3 produces ammonium sulfocyanate:

$$CO_2 + H_2S + 2NH_3 = NH_4 \cdot CNS + 2H_2O$$

But, as we have just seen, it is not sufficient to find a source of organic substances: It is still necessary to liberate oxygen in the gaseous or dissolved state to ensure their oxidation in producing energy. Now, vulcanism does not produce the slightest trace of free oxygen, and for a very good reason.

It is noteworthy that, in 1873, some learned men thought of utilizing solar ultraviolet light in order to produce organic photosyntheses. In a speech given before the Assembly of German Naturalists, that met at Leipzig in 1873, du Bois-Reymond (23) declared: "Where and in what form did life appear for the first time? Was it in the depths of the sea, in the form of animated mud, as the discoveries of Mr. Huxley seem to suggest? Or was it rather the Sunlight which, still containing a greater proportion of ultra-violet rays, gave birth to it in an atmosphere abounding in carbonic acid? Who will ever tell us?"

Moore (24) saw, in the photochemical synthesis of formaldehyde by sunlight in the presence of catalysts, the starting point of the syntheses he sought. P. Becquerel (25) has placed these syntheses: "under the effects," he said, "of solar radiations much more ardent than today and in a very different atmosphere, charged with carbonic acid, ammonia and other vapours emitted by the volcanoes. The ultra-violet radiations have then caused the synthesis of organic substances in the upper atmosphere (as in the experiments of D. Berthelot and of Stoklasa); in this way formaldehyde, sugars and amides are formed."

Becquerel believed the primitive atmosphere to be rich in oxygen and that is why he placed these syntheses in the upper atmosphere. However, the upper atmosphere is very poor in water vapor and only traces of these organic substances could be produced in the gaseous phase. In fact, the near solar ultraviolet radiation of longer wavelength, which brings about their photolysis, is more intense than the far shortwave, which produces their synthesis. Besides, these traces would immediately be oxidized by the ozone present in the stratosphere.

The spectrum of the nocturnal luminescence shows only OH bands and no trace of organic compounds in the upper atmosphere. We shall also see that no asymmetric molecules could originate under these conditions.

Becquerel also believed in universal and eternal life, but he assumed that it appeared, evolved, and disappeared independently on each planet on which it became possible. We shall certainly come to this conception in the present work.

Haldane (26), in a short memoir, thought that these reactions could have originated at the surface of the primitive oceans and he described a warm, salty "soup" in which the first organisms might have appeared.

In 1938, Oparin (27) published an important work on the origin of life. His thesis rejected the previous ideas and he did not appeal to photochemistry. For him, the primitive atmosphere did not contain nitrogen, oxygen, or carbon dioxide, but only hydrocarbons, ammonia, and water vapor. He thought that during the period of incandescence of the globe these hydrocarbons reacted with the water and ammonia and formed ternary and quaternary compounds. It was in this way that a reaction, described by A. Tchitchibabine in 1915, converted acetylene into acetaldehyde, which is capable of combining with ammonia at 300°C in the presence of iron oxide.

$$C_2H_2 + H_2O \rightarrow CH_3COH$$

Primitive organic matter would thus have consisted of alcohols, aldehydes, acids, amines, amides, and other compounds. All these compounds, reacting in an aqueous environment, would have given rise to increasingly complex macromolecules in the course of cooling. There are two fundamental objections to this mechanism from the point of view of the origin of life. These are that none of these reactions leads to the liberation of oxygen and that none gives rise to the dissymetric molecules that are characteristic of living matter. Primitive carbon was not in the form of hydrocarbons but of carbon dioxide. Vulcanism provided much more carbon dioxide than methane and never gave acetylene. The atmospheres of the terrestrial planets, such as Venus and Mars, do not contain oxygen and give no evidence of hydrocarbons, but plenty of carbon dioxide. The present atmospheres of Venus and Mars, which are sterile planets, are representative of that of the primitive earth.

In collaboration with Desguin (28), we proposed, in 1939, a theory of the origin of life based on astronomical observations concerning the nature of the planetary atmospheres, the experiments of D. Berthelot and H. Gaudechon on photosynthesis, the explanation of molecular dissymmetry by Pasteur and the considerations developed by Guye. This is the theory that we shall discuss. The cosmic problem of the formation of organic matter on an inorganic sterile planet is linked intimately with cosmogony and geogony. We must explain simultaneously, the transparency of the primitive atmosphere to the far solar ultraviolet and the photosynthesis of ternary and quaternary compounds, the appearance of dissymmetric molecules, the formation of an atmosphere of free oxygen and ozone, that will permit the controlled oxidation of these substances, and the subsequent emergence of life that will be protected from the effect of abiotic radiations.

Therefore, first of all, we must briefly describe our cosmogonic conceptions concerning the formation of the solar system and of our planet, and study the physicochemical evolution of the earth, that is, the formation of the continents, oceans, and atmosphere. In short, we shall discuss the conditions of the primitive sterile environment. We shall then describe the photochemical synthesis of asymmetric macromolecules, and, finally, we shall try to sketch the passage from geochemistry to cytology, showing how sterile organic matter was able to become organized and give rise to the living world in a more and more complex evolution.

References

1. J. Rostand, Cahiers d'études biologiques," No. 3, p. 51 (1957).
2. R. Esnault-Pelterie, "L'Astronautique." Lahure, Paris (1930).
3. H. Henriet, *Compt. Rend.* **135,** 101 (1902); **138,** 203 (1904).
4. N. R. Dahr and Atma Ram, *Nature* **132,** 819 (1933).
5. H. E. Richter, *Schmidts Jahrb. Ges. Med.* 126 (1865), p. 148 (1870).
6. S. Cloez, *Compt. Rend.* **59,** 37 (1864).
7. G. Müller, *Geochim. Cosmochim. Acta* **4,** 1 (1953).
8. M. Calvin, *Sci. News Letter.* **76,** 359 (1959).
9. B. Nagy, W. G. Meinschein, and D. J. Hennessy, *Ann. N.Y. Acad. Sci.* **93,** 25 (1961).
10. G. Astre, *Ann. Obs. Toulouse* **28,** 59 (1961).
11. W. G. Claus, and B. Nagy, *Nature* **192,** 594 (1961).
12. G. Deflandre, *Compt. Rend.* **254,** 3405 (1962).
13. S. Arrhenius, "L'évolution des Mondes." Beranger, Paris (1910).
14. A. Dauvillier, "La poussière cosmique. Les milieux interplanétaire, interstellaire et intergalactique." Masson, Paris (1961). "Cosmic Dust." Newnes, London (1963).

15. P. Becquerel, *Compt. Rend.* **151,** 86 (1910).
16. V. Henri and V. Moycho, *Compt. Rend.* **158,** 1509 (1914).
17. C. E. Guye, "L'évolution physico-chimique." Chiron, Paris (1922); "Les frontières de la physique et de la biologie." A. Kundig, Geneva (1936).
18. Lecomte du Noüy, "L'Homme et sa destinée." La Colombe, Paris (1948).
19. H. F. Osborn, "The Origin and Evolution of Life." Masson, Paris (1921).
20. J. Constantin, "Origine de la vie sur le Globe." Paris (1923).
21. E. Pflüger, *Arch. ges. Physiol.* **10,** 251 (1875).
22. A. Gautier, *Compt. Rend.* **132,** 938 (1901); **150,** 1564 (1910).
23. E. du Bois-Reymond, "Uber die Grenzen der Naturerkenntniss." Veit, Leipzig (1873).
24. B. Moore, "The Origin and Nature of Life." Willian and Norgate, London (1913); Thornton, London (1930).
25. P. Becquerel, *Astronomie* **38,** 293 (1924).
26. J. B. S. Haldane, *Rationalist Annual* 148 (1928).
27. A. I. Oparin, "The Origin of Life." Macmillan, New York (1938); Dover, New York (1953).
28. E. Desguin and A. Dauvillier, *Compt. Rend.* **208,** 294 (1939); *Rev. Sci.* **89,** 292 (1940); "La genèse de la vie, phase de l'évolution géochimique" Hermann, Paris (1942).

III
COSMOGONY AND GEOGENY: The Origin of the Earth

We have seen that the problem of the origin of life assumes full significance, and its solution appears clearly, only if its cosmic, geochemical, and energy aspects are considered, that is to say, when it is placed in the very general framework of the actual evolution of our planet. In this chapter, we shall show how the restricted cosmogonic problem of the origin of the earth and of the solar system is integrated into the general cosmogonic problem of the evolution of the universe.

The Evolution of the Universe

Cosmogony is based on recent knowledge of the architecture of the universe and on the cosmic abundance, age, and evolution of the chemical elements. Just as Galileo familiarized us with the structure of the solar system and W. Herschel with that of the galaxy, so E. P. Hubble unveiled the architecture of the universe to us. Figures for the cosmic abundance of the chemical elements are known for the galaxy. They are derived from spectrographic work at the surface of the sun and various stellar types, in interstellar matter, nebulae, planets, comets, and meteors. Geochemistry and statistical chemical study of the various classes of meteorites complete these investigations. Finally, mass spectroscopy of primary cosmic rays offers a new method of determining the abundance of the elements in the galaxy.

Discussion of all these data together allows the cosmic abundance of the elements to be known approximately. Moreover, by comparing the concentrations of the isotopes of heavy radioelements and of lead, the age of the elements, forming the solar system, may be estimated. These elements are rapidly destroyed in the galaxy, either by natural radioactivity or by the effect of the cosmic rays, so that the problem, not only of their origin, but also of their persistence, may be posed unambiguously today.

Numerous theories of the method of formation of the elements have been proposed to account for the abundances observed. Some, reviving the spurious and insoluble problems of the "age" of the universe, of its beginning and end, have a metaphysical character which renders them inadmissible. Other theories postulate arbitrary previous states of the matter of the universe.

The problem of Kant, who, following Ovid, wished to make the celestial bodies emerge from an immobile, cold, dark "primitive chaos" of cosmic dust, is purely imaginary: We have no evidence that enables us to believe that the universe has ever presented a different aspect than the one it offers today. The universe is composed of too many elements, galaxies, and stars ever to have been different. There are too many galactic and stellar populations for them to have been able, by virtue of the laws relating to large numbers, to show an average state different from that which they present to us today. On the other hand, G. Lemaitre's theory of the "primitive atom" has not been confirmed, either in the domain of the structure of the metagalaxies or in that of the cosmic rays. It attributed to the universe an age of 1.7 and, later, of 3.4 thousand million years, which is less than the age of our elements and less than the age of the sun and the earth. All stellar and galactic evolution would have been prohibited. This is the "short scale" of the "age" of the universe, and it may be wondered what the universe could have achieved in so brief a time.

Today, the adherents of the short scale attribute an age of 11×10^9 years to the universe, even though certain red stars seem to require 20×10^9 years.

If we assume that the universe is in statistical equilibrium, it is possible to look for a mechanism that will connect the evolution of the chemical elements with that of the stars and galaxies. For this, we must utilize our knowledge of the structure and evolution of the stars, of the architecture and expansion of the universe, as well as the main cosmogonic hypotheses concerning its evolution. Only modern hypotheses that are based on relativity, observed data in astronomy and microphysics, and that are exempt from any metaphysical character, may be retained. However they will still be incomplete since they fail to consider *cosmic electromagnetism* which, through the intermediary of cosmic radiation, will permit a high-quantum generative mechanism to be introduced and which will allow new matter to be produced at the expense of the energy.

The application, attributable to Lord Kelvin, of the kinetic theory of gases to cosmology, in which the stars play the role of the molecules, has been elaborated on by H. Poincaré. The theory maintains that, although the spherical galaxies and globular clusters are in statistical equilibrium, the spirals will represent dynamic configurations. Taking the high stellar densities that are achieved at the center of the clusters into consideration, the theory accounts for stellar collisions, equipartition of energy, and the origin of the rotation of the stars and stellar associations. The formation of the planetary systems represents merely a particular case of stellar association. The theory is independent of the expansion of the universe, which does not relate to the cosmic unit of the metagalaxy. This expansion may be recent and localized in the minute fraction of space which alone is perceptible to us.

One is thus led (1) to a new model of the metagalaxy in statistical equilibrium, in which the evolutions of the elements, stars, and galaxies are closely associated, and in which cosmic radiation plays an essential role. The galactic cycle is maintained by the satellite globular stellar clusters that play many cosmological roles. It is these clusters, derived from intergalactic matter at rest, that ensure that matter will be continually exchanged between the galaxies. The elements are recreated from the hyperdense stars when stellar collisions occur at the center of these clusters at the same time that new giant stars are forming. The cosmic abundance of the elements seem to be a statistical phenomenon.

These rare collisions are the transient "hot points" of the universe in which the temperature may momentarily reach the several thousand million degrees that is necessary for atomic synthesis. It is realized that, in consequence of Stefan's law, no permanent source can exist at such temperatures in the universe.

Electromagnetism, through the intermediary action of cosmic radiation, is also the "winder-key" that counterbalances the quantic degradation of energy and ensures the formation of new matter from the radiation. Thus, the universe does not grow old and the "short scale" of its duration is in harmony with the "long scale."

The Origin of the Solar System

Since Descartes, numerous hypotheses have been advanced to account for the origin of the solar system. The theories, based on the accretion of cosmic dust, to which the names of Kant, Faye, du Ligondès, Berlage, and Weizsäcker, are attached, are inadequate. None of these theories accounts for the structural details of the solar system and all are fundamentally unsound.

Nothing, in fact, is less primitive than the cosmic dust (2). The cosmic dust that exists today in the solar system has been caused by the incessant fragmentation of divided solid matter, asteroids, and comets, and the cosmic dust that appears in the galactic plane has been brought about by the explosions of novas and supernovas. Its formation requires prior organization of atoms, particles, celestial bodies, stars, and planets, which have required a long period of evolution.

Instead of accretion, we see, on the contrary, the asteroids splitting up, the cometary nuclei dispersing in a shower of shooting stars, and stellar explosions giving rise to absorbent clouds of cosmic smoke and to gaseous nebulae. The works of Daubrée, S. Meunier, and the geochemists have shown that the whole idea of accretion must be excluded from the mechanisms that seek to explain the nature of meteorites and planets.

Laplace's theory, and all theories that attribute the formation of the solar system to the spontaneous internal evolution of a protostar or of a gaseous nebula, are no more satisfying. Not only are we unacquainted with a star that would meet Laplace's specifications, but its evolution would have demanded a prohibitive period of time, and this hypothesis also runs into the fatal objection of the angular momentum. The knowledge of celestial mechanics forbids this conception, and the kinetic theory of gases is opposed to the intermittent ejection of Laplace's "rings." Neither the theoretical work of Roche, nor that of Poincaré, have been able to validate this conception. The dynamic evolution of a gaseous mass, in contracted and accelerated rotation, has instead been transposed, by H. Poincaré and Sir J. Jeans, to galactic evolution.

Some astronomers, including T. J. J. See, have tried to include the mutual *capture* of celestial bodies into cosmological doctrine. If this idea is fruitful in accounting for the formation of the double stars and explaining the existence of certain irregular microsatellites, such as Phoebe, it is incompatible with the essential characteristics of the solar system, so strongly stressed by Laplace.

The *electromagnetic* hypotheses, proposed by some physicists in the wake of K. Birkeland, are untenable. The same applies to theories that assume that the solar system was constructed from internal stellar explosions.

An important group of theories based on the causation of the *tides* was inaugurated by Sir G. H. Darwin's famous work on the evolution of the earth-moon system. If his resonance theory encounters fatal objections, as Sir H. Jeffreys' work has shown, the application of the tides' causation to the galaxies, to the close binaries, and to the evolution of some celestial bodies in the solar system has been fruitful. Study of the slowing down of the earth's rotation enables the age of the solar system to be estimated, in agreement with the data of radioactivity.

The cosmogonic hypotheses based, on the causation of the tides by interstellar interactions, held by Bickerton, Jeans, and Russell, also run into objections that have been stressed by L. Spitzer and are likewise based on improbability. The same applies, *a fortiori,* to those based on collisions. Buffon's hypothesis, on the other hand, led to too high a frequency and to too great an abundance of planets, indeed it is notoriously inadequate. The stellar collisions, postulated by Arrhenius, Gifford, and Jeffreys, are even more improbable and lead to the unsatisfactory conclusion that the solar system is, as it were, unique in the galaxy.

In the majority of these hypotheses, the *circularity* of the planetary orbits is sought by the action of a hypothetical "resistant medium" which would have existed at the time the solar system was formed. Not only is this idea numerically inexact but it smacks too much of the will-o-the-wisp to be retained.

The cosmogonic problem thus reaches an impasse. On the one hand, the works of Johnstone Stoney, Bickerton, Kelvin, Jeans, and Jeffreys show that the planetary systems are, without any doubt, the result of stellar interactions and that shearing or semicollisions are the most effective mechanisms of conferring mechanical properties on these systems. On the other hand, these theories are so improbable that we cannot take them into consideration. The expansion of the universe, which is sometimes invoked, has not been able to manifest itself in the galaxy. The problem, therefore, has been incorrectly stated and the solar system was born, four thousand million years ago, *in a place completely different from where it finds itself today.*

Therefore, the restricted cosmogonic problem is more complex than had been supposed. In fact, it has three aspects: astronomical, mechanical, and physicochemical.

The Astronomical Problem

The astronomical problem may be dealt with on the basis of the theory of evolution of the universe and of the galactic model indicated previously. In this model, the evolutions of the chemical elements, stars, and galaxies are closely associated. The stellar globular clusters play an essential cosmological role. Applying the kinetic theory of gases to the metagalaxies, galaxies, and clusters, in accordance with the ideas of Kelvin (assuming that the stars are acting as molecules), it can be shown that stellar interactions are frequent. The work of Faye, du Ligondès, and Poincaré, on the dynamics of an ultrararefied cloud of cosmic dust, has already indicated what the dynamics of a cluster could be. The work of von Zeipel and others has shown that the stellar density at the center of a cluster could be 700,000 times greater than in our galactic region. In fact, they contain many more stars than the single giants perceptible to us, as their spectra and the disparity in the associations of double stars that have emerged from them indicate. The free stellar path is comparable to the diameter of the cluster. The cluster is traversed in a time of the order of a million years, so that it will have been covered thousands of times, and the equipartition of the stellar kinetic energy will be almost completed.

Three types of interstellar interactions can be distinguished: (*a*) pseudoimpacts that only give rise to tidal effects, to which the equipartition is owed. When these effects become considerable, the loss of energy brings about *capture* and associations of double, and then of multiple, stars. The large orbital moment of these double stars and the rapid stellar rotations are the consequences of this. (*b*) Central or face-to-face collisions are catastrophic events accompanied by momentarily very high temperatures and nuclear reactions, by means of which the hyperdense stars form new red giants consisting of the new elements. It is in these collisions that we find the source of the permanent *rejuvenation* of the universe and the source of nuclear and radioactive energy. (*c*) Semi-collisions, bring about capture and are followed by a series of tangential collisions that end in the fusion of both stars into a single star. This is the way the planetary systems have formed.

What is true for the clusters is *a fortiori* true for the center of the galactic nucleus, where the stellar population is even denser. In this way, our solar system originated at the center of the nucleus, from the capture and then the union, of two stars of similar mass. The plane of the solar system passes, in fact, through the center of the galactic nucleus. The probabilities of interaction agree with the abundance of red giants and double stars. From this it may be deduced that the abundance of the planetary systems is of the order of hundredths. They are therefore innumerable in the galaxy.

It will be possible to verify these considerations by means of computers. Preliminary trials carried out with the MANIAC computer at Los Alamos on a plane model containing a hundred stars, with a time scale of a million years, have already demonstrated the possibility of captures. The restricted cosmogonic problem thus again becomes part of the general problem of the evolution of the universe.

The Mechanical Problem

The mechanical problem of the origin of the solar system has been put forward in the form of the theory of the twin planets (3). Just as two stars are needed to form a planetary system, so two twin planets are needed to form a satellite system. The existence of two planetary families, the giants and the dwarfs, is explained by a succession of semi-collisions, followed by final oscillations of the sun, resulting from the union of the two generative K dwarf stars. The orbital moment of the system is easily provided by the latter. The nearly exponential decrease in the amplitude of these effects as a function of time, leads to Bode's law. When the twin planets are of comparable mass, the combination of their coplanar, elliptical, symmetrical, and secant orbits will cause the resultant final planet to revolve on a circular orbit.

The interaction of stars of different temperature, mass, and spectral type may give rise to varied effects, for example, to planetary systems in which the orbits are very eccentric or almost cometary, and in which the planets will have no satellites.

The union of twin planets may give rise to events of three types: (*a*) In the case of the giants, coming together at the end of several millennia, still in the gaseous state, *capture* with successive semicollisions has occurred in the case of Saturn. This explains the structural similarity between the planetary system and its satellite system. In fact, Saturn possesses two families of normal satellites, the giants, Titan and Japet, and the dwarfs. In this way, the discontinuities which we observe in the distances and masses of the planets and satellites are explained. In the case of Jupiter and Uranus, the two planets have united directly and have given rise to only one family of dwarfs. In this way, the *differential rotations* of the sun and giant planets are explained. (*b*) In the case of the terrestrial planets, whose twins have united after several centuries in the molten state, normal satellites could not be produced. But in the exceptional case of the earth, the two twins have united tangentially, bringing a considerable angular moment that has led to the *piroid* of Poincaré and to the earth-moon system, a *double planet* which has evolved in accordance with Darwin's theory. (*c*) Finally, in the case of the two twins which must have formed Olbers's planet, postulated to exist between Mars and Jupiter, the planets were so small that they were captured in the solid state, at the end of their *chemical and mineralogical evolution*. They were progressively distributed along the resultant circular orbit, giving rise to the asteroid rings.

The genesis of Saturn's ring was similar, since the first two twin satellites of this planet were the smallest of the solar system and, for this reason, they were captured in the solid state. Thus, contrary to classical ideas, which see the asteroid rings and Saturn's ring as a gestating planet and satellite in accretion, we shall regard them as the debris of a planet and of a satellite.

The Physicochemical Problem

Ignored in all previous (1942) cosmogonic hypotheses, the physicochemical problem is not the least important. It is dominated by the cosmic abundance of the elements, the nature of Russell's solar mixture, and the kinetic theory of gases, applied, this time, to the proper object. It must account for the very different chemical compositions of the sun, giant and dwarf planets, and meteorites. If one wishes to extract all planetary matter from the sun on one occasion, as Jeans did, one would be faced with Spitzer's objections: the matter would explode. If it is extracted in insufficient "quanta," as Chamberlin and Moulton wished, one is faced with objections taken from kinetic theory: The gas would evaporate in space. But all planets and all satellites are nonetheless formed at the cost of a considerable loss of matter. For each temperature, the theory assigns a critical mass, and chemical differentiation is the result of a fractional thermal evaporation. In this way, it is possible to give theoretical bases to geochemistry.

The surface layers of the sun seem to be formed of a gaseous mixture, the so-called "Russell mixture," conforming to a considerable extent to the cosmic abundance of the elements, that is to say, formed essentially of helium and hydrogen together with a small proportion of elements forming the oxygen, silicon, and iron groups. The cosmic abundance of these groups can be schematized as follows: $H = 70$, $O + N + C +$ etc. $= 0.7$, $He = 29$, $Fe + Si + Mg +$ etc. $= 0.3$.

Among the numerous types of "Russell mixture" proposed, the mean composition: $H = 50$, $O + N + C +$ etc. $= 6$, $He = 42$, $Fe + Si + Mg$ $+$ etc. $= 2$ can be selected.

The sun contains less hydrogen and more helium on account of its advanced age, 4.5×10^9 years, and the thermonuclear cycles which have maintained its radiation. The solar system is composed of this type of mixture for the isotope composition is uniform everywhere with the exception of geological and biogenic segregations. It is possible to show how the giant planets, because of their great mass, have a composition close to that of this mixture, since they only lost the major part of the helium and uncombined hydrogen during their stage as gaseous protoplanets at high temperature. So, for the most part, they are composed of light elements forming water ice, methane, and ammonia.

It has been quite different with the terrestrial dwarf planets. The application of the kinetic theory of gases, consideration of the critical molecular escape velocities, spectrographic study of the solar atmosphere, thermochemistry, and the laws of chemical equilibrium enable the nature of the fractional molecular evaporation undergone by the gaseous terrestrial protoplanets to be made explicit. Apart from rare endothermal molecules, such as O_3 and C_2H_2, only biatomic radicals exist at the temperature of the sunspots (4000°K). Hydrogen will not be retained, either in the atomic state or in the molecular state, or in the form of light hydrides, CH, NH, and OH, that are present in the spots, but only in the form of rare heavy hydrides, MgH, AlH, SiH, CaH, which are present in small amounts. This explains how the earth has retained only 2×10^{-5} of hydrogen, which forms the hydrosphere.

The light biogenic elements, C, N, and O, also escape in the atomic and molecular state and in the form of light CH, NH, OH, and CO radicals. They are retained only by their chemical affinities, that is, as carbides, nitrides, and heavy oxides, SiO, MgO, AlO, TiO, and ZrO, present in the sunspots. They are mainly retained, therefore, via silicon and earth metals. Since the affinity of oxygen is much greater than that of carbon and nitrogen, the greater part of this element has been retained. It is the most abundant terrestrial element. All the electropositive elements have been completely oxidized.

Fluorine is also retained in the form of heavy fluorides MgF, SiF, and SrF that are present in the spots. Neon escapes almost completely, but all the heavier elements have been preserved.

A good many theoreticians have believed, following Legendre, and on the strength of theories of cosmic dust accretion, that all the heavenly bodies were made from the same material and that their different densities are the result of compression effects. But these views, supported by Jeffreys (4), Ramsey (5), and Kuhn and Rittmann (6), are completely opposed to what we know of the cosmic abundance of the elements. In fact, the density, flatness and moment of inertia of the globe are in harmony with Roche's model (1881). Daubrée's work has shown how the terrestrial planets were formed of a nucleus of cosmic iron, of the siderite kind, surrounded by a scoria of peridotite rock, of the stony meteorite type.

References

1. A. Dauvillier, *Compt. Rend.* **237**, 1298 (1953); *Rev. Sci.* **91**, 90 (1953); "Cosmologie et chimie." Presses Universitaires de France, Paris (1955); "Cosmogonic Hypotheses. Theories of Cosmic Cycles and Twin Planets." Masson, Paris (1963).
2. A. Dauvillier, "Cosmic Dust." Newnes International Monographs on Astronautics and Astronomy, London (1963).
3. A. Dauvillier, *Compt. Rend.* 786 (1942); "Genèse, nature et évolution des planètes." Hermann, Paris (1947); "L'origine des planètes" Presses Universitaires de France, Paris (1956).
4. H. Jeffreys, "The Earth," Cambridge Univ. Press, London and New York (1924, 1929, 1952).
5. W. H. Ramsey, *Monthly Notices Roy. Astron. Soc.* **108**, 406–413 (1948).
6. W. Kuhn, and A. Rittmann, *Geol. Rundschau* **32**, 215 (1941).

IV
THE ORIGIN OF THE CONTINENTS, OCEANS, AND ATMOSPHERES

In this chapter, we are going to retrace the main features of the earth's primitive cosmic chemistry, which resulted in the formation of its geological structures, oceans, and atmospheres. In the preceding chapter, we tried to lay down the theoretical bases of geochemistry, by showing how the terrestrial planets, and particularly the earth-moon system, had lost a considerable amount of matter during their condensation: All free hydrogen, helium, and the major part of the light biogenic elements were lost, with the exception of oxygen, which was retained in the form of metal oxides and ozone. The terrestrial planets have appeared to us as *cosmic ashes* resulting from the combustion of the elements that made up "Russell's solar mixture." We are going to retrace the main features of geochemical evolution as a function of the decreasing surface temperature during surface cooling.

The Cosmic Chemistry of the Lithosphere

We have seen how, at about 3000°C, the central ferronickel core had been condensed by a change of state in the form of *rains,* from the gaseous phase, whereas the oxides, which were to form the lithosphere, remained in the gaseous state. The earth, molten and incandescent, was at this epoch only about the volume of its core, that is, it had a radius of 5000 km only.

This ferronickel core was surrounded by a scoria and by a primitive atmosphere of silica and hydrocarbons. Acetylene, C_2H_2, and hydrogen silicide, SiH_4, were formed by direct synthesis. They will be stable and abundant at this temperature: It was this acetylene synthesis that Berthelot achieved in the electric arc. The primitive endothermal acetylene was able to institute a source of hydrogen by cooling and spontaneous decomposition (cracking):

$$C_2H_2 \rightarrow C_2 + H_2$$

The acetylene could polymerize to form benzene, C_6H_6, styrolene, C_8H_8, naphthalene, $C_{10}H_8$ and to give cyano compounds in the presence of nitrogen and under the influence of lighting:

$$C_2H_2 + N_2 \rightarrow 2\,HCN$$

But although these gases are easily formed by synthesis in the spark and in the arc, cyanogen, C_2N_2, is dissociated at these temperatures. Cyano compounds will be formed indirectly, as we shall see.

The primitive atmosphere of silica was formed by reactions such as

$$SiO + O_3 \rightarrow SiO_2 + O_2$$

and by combustion of SiH_4.

Moissan (1), showed, by means of the arc furnace, the chemistry at these temperatures: It was the chemistry of the endothermic and stable hydrides, nitrides, carbides, and silicides. So it is seen that titanium, an abundant element, in the presence of carbon, nitrogen, and oxygen, will first give a carbide TiC (F. 3420°K), that will be partially dissociated when the temperature falls, and will be replaced by a yellow nitride Ti_2N_2 (F. 2925°C), and finally by the blue oxide TiO_2 (F. 1600°C). We have shown that it was because of the formation of such compounds that a small fraction of carbon and nitrogen was retained by the earth.

Generally, the terrestrial elements have subsequently combined with each other in accordance with their mutual affinities, and have released the maximum amount of heat. The order of their reactions is indicated to us by thermochemistry. From Le Chatelier's principle, compounds formed by a reduction of temperature will be exothermic compounds and minerals. First of all rains of carbides will be formed such as CFe_3, CMn_3, C_3Al_4, and C_2Ca; silicides such as $FeSi_6$, SiFe, $SiFe_2$, SiC, Si_2Ca, and $SiMg_2$; borides such as NB and TiB; nitrides such as Fe_4N, FeN, N_2Al_2, N_2Mg_3, N_4Si_3, and N_2Ca_3; and hydrides such as CaH_2 and MgH_2; then later, when the temperature has fallen, oxides will be formed such as SiO_2, CaO, MgO, and TiO_2. Therefore carbides such as cohenite, $(Fe, Ni, Co)_3$ C, TiC, carborundum SiC, and nitrides, TiN and FeN are found in meteorites. These carbides are endothermic compounds that only occur and remain stable at a high temperature. Their presence in meteorites is, therefore, of profound cosmogonic significance. Phosphorus and sulfur will produce the phosphides and sulfides that are contained in the stony meteorites, schreibersite (Fe, Ni, Co)$_3$P, FeP, P_2Mg, P_2Ca; troilite or pyrrhotite, FeS; daubréelite, $FeS \cdot Cr_2S_3$; pyrite, FeS_2; and oldhamite, CaS. Generally, the alkalies will fix metalloids such as N, P, S, and As. The alkaline elements will combine with the halogens, Cl and F, and will give volatile chlorides and fluorides that will remain in the atmosphere. These reactions explain the total diffusion of such *dispersed* elements as boron, sodium, iron, and uranium. To these Li, Sc, Cu, Ga, Ge, Br, Rb, Y, In, I, Cs, Re, and Au may be added. All the biogenic elements will be fixed chemically and all the electropositive elements will be completely oxidized or halogenated. The terrestrial or cosmic rocks appear to be *combustion* products, and no trace of free electropositive elements is able to be detected. However, a number of chemically inert or rather inactive elements, such as the rare gases and elements of the platinum family, will remain in the free state. These will follow the ferronickel core as their presence in siderites shows.

However, at the surface of the globe, some thirty elements are found in the free state. Some of these elements form native metals, but the majority are the result of reactions which have liberated them. Volcanoes release hydrogen, the "cosmic nitrogen" of C. Moureu and A. Lepape (that is, nitrogen accompanied by the rare gases except for helium) sulfur and traces of halogens, Cl, F, and I. Free oxygen is an exclusive product of the biosphere. Some metalloids, C, As, Sb, Se, and Te, exist in the free state; so do numerous metals, "cosmic iron" or alloys of Fe, Ni, Co; platinum ore, Ru, Rh, Pd, Os, Ir, and Pt; the gold group, Zn, Cu, and Ag; and the mercury group, Sn, Ag, Pb, and Bi.

When volatile oxides, such as silica, magnesia, limestone, and alumina, condensed on the *scoria* of hydrides, nitrides, carbides, and silicides surrounding the ferronickel core, silicate chemistry began and the silicates entered into reaction with the core to produce the *magma* (2). Following these reactions, the surface sial magma will be in hydrothermal fusion whereas the deep sima will remain in dry fusion.

The silicates that formed at about 2000°C accomplished this by the igneous route exclusively, since water vapor did not exist. Blast-furnace slag melts at about 1850°C.

The primitive silicates of the lithosphere, which no longer exist at the globe's surface, except when they are released from deep eruptions, are those that form the moon and are brought to us by the stony meteorites. Silica will react first of all on the ferruginous scoria to produce dense, fusible ferrous silicates such as the black peridot or fayalite, $SiO_2 \cdot 2FeO$ ($\rho = 4.14$, F. 1205°C), of the volcanic massif of the Azores, and forsterite, $SiO_2 \cdot 2MgO$ ($\rho = 3.19$, F. 1890°C). The principal silicates common to the stony meteorites and to the earth are: peridot,

$SiO_2MgOFeO$ ($\rho = 3.35$), pyroxene, $SiO_2(Mg,Fe)O$ ($\rho = 3.2$), enstatite, SiO_2MgO ($\rho = 3.3$, F: 1400°C), anorthite feldspar: $2SiO_2 \cdot CaO \cdot Al_2O_3$ ($\rho = 2.7$), amphibole, and hornblende, which are ferromagnesium silicates containing some calcium and aluminum, and labradorite, alumino-sodium calcium silicate.

First, the various envelopes of the planet, core, scoria, sima and sial, are arranged in order of density. This fact has even been structured into a "law" by de Launay (3) and has given rise to many misconceptions. Thus, it was thought that the heavy radioelements were concentrated at the center of the globe, when, in fact they are located at its surface. Second, we must take into account chemical analogies and segregations. For example, the close relationship that exists between "congenital" elements such as Sn and W, V, and U, and Ag and Pb, is well known. The vertical distribution of the mineral deposits depends much more on solubility in the magmas, isomorphism, chemical relationship, volatility, and vapor pressures than on gravity.

Nevertheless, the earth's first obsidian ($\rho = 2.4$) vitreous crust, formed by light, poorly fusible, acid, alumino-sodium silicates of the tektite type (A. Lepape), and not, as too many authors thought, by contemporary rocks such as gneiss and granites, which are the result of a long mineralogical evolution, and which appeared later when water vapor formed during later geological epochs.

The Pyrogenic Syntheses of Organic Compounds

Some organic compounds that are stable at high temperatures were first formed in this period of geochemical evolution. We have seen how benzene nuclei were formed from acetylene. Cyano compounds would have been able to originate from ammonia, which may be synthesized directly under pressure at high temperature:

$$2\,NH_3 + C \rightarrow NH_4CN + H_2$$
$$CO + NH_3 \rightarrow HCN + H_2O$$

But this mechanism is very unlikely, since ammonia is too light a gas for the incandescent earth to retain. For the same reason it is not very likely that the cyano compounds originated from nitrogen.

$$C_2Ba + N_2 \rightarrow Ba(CN)_2$$
$$C_2Ca + N_2 \rightarrow CN_2Ca + C$$

Let us also mention Fischer's reaction:

$$BaO + 3\,C + N_2 \rightarrow Ba(CN)_2 + CO$$

But these compounds could have been produced by nitrides and cyanamides, and the carbon could have been retained in the form of "cosmic soot":

$$N_2Ca_3 + 5\,C \rightarrow CN_2Ca + 2\,CaC_2$$
$$CN_2Ca + 2\,NaCl + C \rightarrow 2\,NaCN + CaCl_2$$

Formic acid could have been produced according to Berthelot's reaction:

$$CO_2 + H_2 \rightarrow HCO_2H$$

The acetylenic hydrocarbons and cyanogen have led, by pyrogenic syntheses, to cyclic hydrocarbons, such as benzene, and to a number of heterocyclic compounds, such as

thiophene (C_4H_4S: HC=CH–CH=CH ring closed by S)

pyrrole (C_4H_4NH: HC=CH–CH=CH ring closed by NH)

and pyrazole and pyridine.

Thiophene is produced by the action of sulfur on acetylene and pyrrole by the action of ammonia.

Pyrazole results from the reaction between acetylene and diazomethane:

$$N_2CH_2 + C_2H_2 = C_3H_3N_2H \text{ (pyrazole)}$$

Pyridine is produced by the action of acetylene on hydrogen cyanide:

$$HCN + 2\,C_2H_2 = C_5H_5N \text{ (pyridine)}$$

Gautier (4) reported the presence of sulfocyanates in volcanic exhalations:

$$CO_2 + H_2S + 2\,NH_3 = SCN \cdot NH_4 + 2\,H_2O,$$

but these do not contain the methylquinoleinic compounds that are encountered in mineral oils.

Ammonia and carbon dioxide combine to form isocyanic acid:

$$CO_2 + NH_3 \longrightarrow CONH + H_2O,$$

which can fix a second NH_3 molecule and give urea:

$$CONH + NH_3 \longrightarrow CO(NH_2)_2$$

It is also possible to get urea via ammonium carbamate and ammonium isocyanate:

$$CO_2 + 2\,NH_3 \longrightarrow CO\begin{matrix}\diagup NH_2\\ \diagdown O\cdot NH_4\end{matrix} \longrightarrow H_2O + CON\cdot NH_4 \longrightarrow CO\begin{matrix}\diagup NH_2\\ \diagdown NH_2\end{matrix}$$

Urea may condense into purine and pyrimidine nuclei and into imidazole, which are constituents of nucleoproteins.

Ammonia, reacting at red-heat with carbon disulfide, gives H_2S and sulfocyanhydric acid. With ammonium carbamate at 160°C, ammonium thiocyanate, NH_4CSN, and thiourea, $(NH_2)_2CS$, are again produced. Amines react at red-heat with CS_2, and decompose into carbon, H_2S, and thiocyanic acid. All these reactions of cosmic paleovulcanism are now unable to occur in the presence of an atmosphere of oxygen.

On the primitive earth, these heterocyclic compounds served to build the nucleoproteins, but they remained occluded in the lithospheres of Mercury, Mars, the Moon, and Olbers' planet.

It has been shown by the theory of the twin planets how the two twin planets of Olbers had eventually come to interact at the end of their chemical and mineralogical evolution, which gave rise to the complex mineralogy of meteorites. In this interaction, the source of all the finely divided solid matter existing in the solar system was found: asteroids, captured microsatellites, such as those of Mars, cometary nuclei, meteors, meteorites, and cosmic dust. Therefore, it is not surprising to observe simultaneously in many comets the spectrum of sodium and that of CN. A powerful argument has been proposed in support of this hypothesis by the discovery, by Calvin (5), of heterocyclic compounds of the purine and pyridine type in the interior of a stony meteorite. It may be thought (6) that these heterocyclic substances, that have been decomposed by heat and solar radiations at the surface of Mercury and the moon, still exist at the surface of Mars and that their partial carbonization produces the darkening that can be observed in the seas there. Photochemical syntheses of the same type as those of Berthelot and Gaudechon cannot, indeed, occur in the gaseous phase from CO_2 and H_2O, as the absence of oxygen in Mars' atmosphere shows. Instead of designating the substances responsible for the CH bond recognized by W. Sinton as "living Martian creatures" we shall call them, more simply, heterocyclic pyrogenic compounds.

The Formation of Water Vapor and of Carbon Dioxide. The Origin of the Oceans

The problem of the origin of water, that is to say, of the hydrosphere, is inseparable from geochemical evolution. A great many authors, in trying to outline the cosmic chemistry of the Globe, have thought that the two substances, water vapor and carbon dioxide—which must have played an essential role in geochemical evolution—were produced directly in the primitive atmosphere, or even that they had an extraterrestrial origin and were captured by the earth in interstellar space. They forgot that these substances do not exist at high temperatures or in the presence of ultraviolet light, and that they could only have had a delayed and very indirect origin.

Water and carbon dioxide can be dissociated at temperatures greater than 1500°C and it is incorrect to assume that they existed in the cosmic period that we have considered so far. Figure IV-1 shows, for

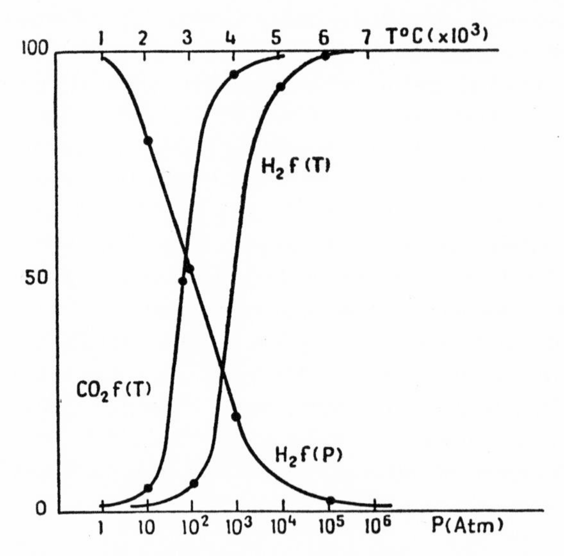

Fig. IV-1. Dissociation of hydrogen and carbon dioxide as function of temperature (at 1 atmosphere) and pressure (at 6000°K).

example, the dissociation curves of molecular hydrogen, H_2, and carbon dioxide, ($CO_2 \rightleftarrows CO + O$), as a function of temperature (1000 to 6000°C) at atmospheric pressure, and of hydrogen at 6000°C as a function of pressure (from 1 to 10^6 atm). It is seen that the dissociation of CO_2 begins at about 1000°C, that it is about 50% at 2700°C, and complete at 5000°C. That of water is similar. It begins at 1200°C (0.034%) and reaches 13% at 1 atmosphere pressure. The dissociation $H_2O \rightarrow OH + H$, like that of CO_2, may also be effected under the influence of ultraviolet light, by wavelengths shorter than 1850 Å.

We shall attribute the water and carbon dioxide which are the main products of present day vulcanism, to the chemical evolution of the primitive surface magma during its cosmic cooling, that is to say, to a phase of *cosmic paleovulcanism.* This magma, incandescent glass in dry fusion that had the composition of stony meteorites, was rich in iron oxide. Chaudron (7) showed that the only oxide stable below 1370°C was the sesqui-oxide, Fe_2O_3. It can be reduced by H_2 and CO above 600°C, forming water vapor and magnetite:

$$3\,Fe_2O_3 + H_2 = 2\,(FeO{\cdot}Fe_2O_3) + H_2O$$

This hydrogen, which no longer existed in the free state in the primitive atmosphere, existed in abundance in the dissolved state, in the ferronickel core; and occluded, in the vitreous magma, as well as in the form of dissociable metallic hydrides. Graham (8), comparing the gases occluded in siderites to the gases occluded in the native terrestrial metals (Fe, Pt, Au) and in industrial metals, showed that Lenarto's iron ($\rho = 7.79$, Fe: 90.88, Ni: 8.45, Co: 0.66, Cu: 0.002) gave up, *in vacuo,* 2.85 times its volume of gases consisting mainly of hydrogen (86%), with traces of N_2 and CO. On the other hand, industrial iron releases mainly CO (2.66 volumes, CO: 50.3, H_2: 35, CO_2: 7.7, N_2: 7), and never absorbs more than its own volume of hydrogen. This observation has been confirmed by Wright (9) who found hydrogen concentrations varying from 18.2 to 76.8% in the gases extracted from siderites.

Magnetite and chromite (Fe, Mg)O (Cr, Al)$_2$O$_3$, $\rho = 4.3$–4.5) are some constituents of meteorites. Magnetite may also result from the oxidation of ferrous oxide by water vapor. Meunier (10) found, in the Kiowa (Kansas) meteorite, some magnetite which he reproduced by heating a fragment of meteorite to red-heat in water vapor. The reaction was as follows:

$$3\,FeO + H_2O \rightarrow FeO\,Fe_2O_3 + H_2$$

"It must therefore be concluded," he said, "that the oxidation came about in the meteoric environment itself where steam jets, analogous to our terrestrial soufflards, broke through into contact with the previously formed, redhot metallic rock."

Magnetite and chromite accompany the deep peridotite rocks such as dunite. The chromite chimneys of the eruptive massif of Southern New Caledonia are well known. Chromite and serpentine, $2SiO_2{\cdot}3MgO{\cdot}2H_2O$ ($\rho = 2.67$) are produced by the weathering of peridot. The ferrous oxide of the pyroxenes, diopside, amphibole, peridot, olivine and fayalith thus seem to us to be a *reduction* product.

The emergence of magnetite and ferromagnetic minerals is of extreme importance because these microcrystals will be magnetized in the cosmic magnetic fields prevailing at the level of the globe, when the cooling has brought them to their Curie point, which is some 580°C for magnetite. The surface rocks of the globe that remain below this critical temperature will receive a long-lasting magnetization which will be the origin of the geomagnetism (11), which seems to be a residual thermomagnetism. Magnetite is abundant in basalts and its deposits sometimes form true mountains. The deep basalt containing the grains of ferronickel like those of Orifak and meteorites plays a major role in geomagnetism.

As for water vapor at high temperature, it played an essential role in the chemical evolution of the lithosphere before it condensed to form the oceans. The oceans represent the *exudate from the young earth during the elaboration of the lithosphere.* The hypercritical water in the deep magmas did not exist in the dissociated state (it did not contain free oxygen) but in the form of mixed silica-water solutions. The water was bound in a labile form to amorphous silica compounds. The solid eruptive rocks, dried at 100°C, still contain, on average, 1.15% of water. The stony meteorites contain 0.5% of water. This water will react, in its turn, with ferrous oxides to give magnetite and hydrogen again and on all the carbides, hydrocarbons, nitrides, hydrides, silicides, phosphides and sulfides of the magma, giving rise to the *chemistry of vulcanism.* This chemistry is very complex but is characterized by the production of reducing compounds and combustible gases, to the exclusion of free oxygen and oxidizing compounds.

The action of water on the metallic carbides is particularly important because in that the indirect origin of the, primitive, young, volcanic carbon dioxide nourishing the biosphere will be found. In 1854, G. Forchhammer showed that the native iron of Nakorniak contained iron and nickel carbides. The basalts of Disko, which are considered to be samples of sima, contain more iron carbide than many meteorites. Gautier proposed the following reactions to explain the production of methane and carbon monoxide:

$$Fe_2C + 2\,H_2O = 2\,FeO + CH_4$$
$$3\,Fe_2C + Fe_2O_3 = 3\,CO + 8\,Fe$$

However, many carbides cannot be decomposed by water vapor. These include Fe_3C, C_2Cr_3, C_2Cr_4, CMo_2, CW, CV, CZr, and CTi. On the other hand, Mn_3C, C_2Th, C_3Al_4, C_3U_2, as well as the alkali earth and rare earth carbides may be decomposed, with the formation of gaseous, liquid, and solid hydrocarbons (CH_4, C_2H_2, C_2H_4). Moissan showed that 4 kg of C_3U_2 gave 100 g of liquid hydrogen carbides.

The last manifestations of vulcanism are accompanied by the production of asphalt and bitumen. The carbonaceous meteorites are the result of reactions of this type. Certain meteorites contain up to 10% of carbon, and the action of water on iron and manganese carbides containing this amount of carbon gives methane and liquids with the odor of mineral oil. Let us recall the theories of Berthelot (1869), Mendelejeff (1877), Abich (1879), and Moissan (1896) regarding a similar mineral origin of the petroleums. Yet the *optical activity* of petroleum, that was discovered by Biot (1835) and explained by Pasteur (1884), the sedimentary position, and the chemical nature imply a *biogenic* origin.

However, some mineral oils are of volcanic origin. Thus, there exists at Trinity, in the English Antilles, an ancient circular crater, a permanent, worked source of an asphaltic black wax. It is the Asphalt Lake, with a surface area of some 46 hectare.

Moissan's carbides give rise to the reactions:

$$2\ C_3Al_4 + 12\ H_2O = 4\ Al_2O_3 + 6\ CH_4$$
$$C_2Ca + H_2O = CaO + C_2H_2$$

The metals are oxidized and hydrogen carbides appear. The latter react with the water vapor at high temperature according to Brodie's reaction:

$$CH_4 + H_2O = CO + 3\ H_2 \qquad -50.7 \text{ cal.}$$
$$CH_4 + 2\ H_2O = CO_2 + 4\ H_2 \qquad -41.2 \text{ cal.}$$

Carbon monoxide is the result of other reactions. We have seen that the primitive atmosphere of the globe contained ozone and acetylene produced by direct synthesis. The fall in temperature will render these endothermic compounds unstable and will allow them to react:

$$C_2H_2 + O_3 = 2\ CO + H_2O$$

Carbon monoxide has played an important geochemical role. It may possibly have formed metallic carbonyls, but it is certainly responsible for the formation of graphite and diamond. Black or transparent diamonds, smaller than a millimeter in size, were discovered by Foote in 1891 in the siderite of Devil's Canyon (Fe: 91–95, Ni: 1–7) which also contained carborundum and graphite. This discovery was confirmed by Mallard, Friedel, and Moissan. It showed that the siderite had crystallized at high temperature and pressure.

On our planet, diamond accompanies the deep peridotite rocks, which are rich in garnet, ilmenite, magnetite, and platinum metals. It is found in the Southern part of Africa, in the diatremes or exploded or perforated volcanic chimneys, in a blue, serpentine clay. Daubrée said it had a volcanic origin and held it to be abundant in the deep magma. Goldschmidt observed that diamonds contain inclusions of iron and titanium oxides and suggested the reaction:

$$CO + Fe \rightarrow C + FeO$$

which would take place at 1800°C and 18,000 atmospheres of CO. But this pressure is insufficient to form diamonds.

The synthesis of small octahedra was achieved by Moissan in 1894 and then, recently, at the Gal Electric Co., where 2-mm crystals have been obtained in 16 hours at a temperature of 2800°C and a pressure of 10^5 atmospheres, corresponding to a depth of 400 km.

As Le Chatelier showed, the carbon-oxygen system gave, above 3000°C, a stable compound, the sesqui-oxide of carbon C_2O_3, which dissociated on cooling into CO_2 and CO. The latter is stable above 1000°C and itself dissociates at a lower temperature into carbon and carbon dioxide; this is Deville's reaction:

$$2\,CO \rightleftharpoons C + CO_2 \qquad +39 \text{ cal.}$$

The gaseous mixture contains, for example, 99% of CO at 1000°C and 99% of CO_2 at 400°C. At 700°C the proportions of the two gases are largely the same.

Carbon monoxide may also react with hydrogen:

$$2\,CO + 2\,H_2 \rightleftharpoons CO_2 + CH_4$$

This reaction, occurring when the gas volume is reduced, is favored by increased pressure. At 1 atmosphere, equilibrium occurs at about 600°C. At 1000°C there is only carbon dioxide, and at 300°C, only methane. The latter is still produced according to Brodie's reaction. Carbon monoxide also reacts with water vapor:

$$CO + H_2O \rightleftharpoons CO_2 + H_2$$

This reaction is quite comparable to that of blast furnaces during the reduction of ferric oxide by carbon monoxide (Boudouard):

$$3\,CO + Fe_2O_3 \rightleftharpoons 3\,CO_2 + 2\,Fe$$

The reactions to which we are referring are, therefore, the very simple ones of heavy chemical industry. In any case, CO is soon replaced by carbon dioxide. Generally, the most stable state of the ternary mixture of C, H, and O corresponds to the formation of methane, carbon dioxide, and water. Moissan showed, in 1901, that the terrestrial carbon was originally contained in the metal carbides, to the extent that these carbides could be decomposed by water vapor.

Did the water and carbon dioxide appear directly when the primitive magma consolidated or have they been produced gradually, in the course of the geological periods, by the activity of vulcanism?

Suess thought that the water of the oceans had been formed progressively. We think, rather, that the major part of the oceanic waters are primitive. The recognized geological epochs have lasted only 500 million years, whereas the reactions to which we are referring took place shortly after the moon broke away, 4.5 thousand million years ago. If the oceans had been formed gradually, their initial salinity would have been so high that life would never have been able to emerge.

Wieliezka's rock-salt, which contains inclusions of hydrogen, methane, and carbon monoxide, is perhaps a sample of these primitive halides. Our thesis will therefore be that the oceanic waters appeared directly, on one occasion only, and that the water that issues from the volcanoes at the present time is cyclic and not juvenile. This could not be true for carbon dioxide. Geochemistry shows us that the CO_2 fixed in the calcareous sediments of biogenic and marine origin is 400 times more abundant than all the CO_2 of the ocean, atmosphere, and biosphere combined. If all this carbon dioxide had been liberated at the same time as the water, the oceans would have been too acid for life to have been able to emerge. Life would have been asphyxiated as soon as it appeared.

But the oceans do not represent all the water of our planet. Because of the internal radioactive heat, the cooling involves only the superficial sial, that is, a thickness of 200 km, and if the latter still contains 1% of water, that represents three times that of the oceans. This mass of water would have required a mass of hydrogen 9 times smaller, that is, $\frac{1}{2} \times 10^{24}$ g, or an average content of combined and occluded hydrogen of 2×10^{-3} in the primitive sial. This low concentration has probably been exhausted in producing all the water on the planet. On the other hand, this water requires the occurrence of a mass of magnetite 14 times greater, that is, 14% in the sial. The surface rocks contain hardly more than 5%, but the lavas and oceanic subsoil contain large amounts (12.7%). The lavas of Etna contain up to 25% of iron oxide.

Furthermore, the carbon dioxide could never have formed an abundant primitive atmosphere. Boussingault showed in 1833 that the very small amount of CO_2 in equilibrium in the atmosphere (2.6 to 3.4×10^{-4} or 2.50 m of normal CO_2) represented the balance between the young CO_2 of volcanic origin and its absorption by the ocean, biosphere, and alumino-sodium silicates.

In 1878, T. Schloesing showed the regulatory role of the oceanic waters in this equilibrium. The carbon dioxide fixed in sea water in the form of calcium bicarbonate, $CaHCO_3$, namely, 6.55×10^{19} g, is 27 times more abundant than the CO_2, 2.43×10^{18} g, existing in the air. When the pressure of the atmospheric CO_2 alters from 10^{-4} atm to 5×10^{-4} atm the amount of CO_2 fixed in the ocean alters from 4.57×10^{19} g to 7.36×10^{19} g.

The CO_2 cycle is a very active one. Produced by vulcanism, it is fixed by the alumino-sodium silicates and the biosphere and converted into alkaline carbonates and sedimental calcareous rocks. The biosphere has never risked being "asphyxiated" by vulcanism.

If the primary flora of our temperate regions was so luxuriant, it is not because the atmosphere was rich in CO_2 but rather because of a tropical climate.

A large part of the carbon dioxide from the magma remains fixed in the rocks. So it is that 1 km^3 of granite contains 9×10^{14} g of CO_2 in the fossil state. All the CO_2 released by vulcanism is no longer young. When a lava flow recovers a calcareous deposit, the latter is dissociated by the heat and releases its carbon dioxide.

This mechanism was brought to light by A. G. Högbom in 1894. He showed that the young volcanic carbon dioxide was, at the time of its liberation, absorbed by the biosphere and immediately fossilized in the form of calcareous deposits, such as chalk and corals.

If we summarize carbon's cosmic history, we see it first of all, in a stellar state in the atomic form C, and then as cosmic soot in the molecular form C_3, C_n. It then forms the CH radical, which is polymerized into C_2H_2 and other more complex molecules. In a protoplanet, it forms metal carbides which are decomposed into hydrocarbons as soon as water vapor appears. These hydrocarbons are, in their turn, decomposed by H_2O into CO, which, being unstable at low temperature, gives rise to volcanic CO_2. This volcanic CO_2 is absorbed by the biosphere and gives rise to the organic macromolecules of living matter. Finally, after the death of the calcareous organisms, it is fossilized in marine deposits, chalk, corals, and dolomite, and partially escapes the carbon cycle.

This mechanism justifies J. Pompecky's phrase (12, 13) "Ohne Radium, kein Vulkanismus; ohne Vulkanismus kein Leben, ohne Radium kein Leben mehr auf unserem Planeten." Many authors have suggested different reactions. Thus Russell (14), considered the abundance of methane on the giant planets, and thought that it resulted from the reaction:

$$CO_2 + 4\,H_2 \rightarrow CH_4 + 2\,H_2O$$

Oparin (15) also thought that the primitive atmosphere of the globe, after the oceans had condensed, consisted not of carbon dioxide, but of hydrocarbons. He invoked the reaction:

$$2CH_4 \rightarrow C_2H_2 + 3H_2 \qquad - 91 \text{ cal.}$$

and thought that these organic compounds would have been able to arise through Chichibabin's (1915) reaction:

$$C_2H_2 + H_2O = CH_3COH \text{ (acetaldehyde).}$$

We believe, on the contrary, that carbon dioxide followed the hydrocarbons, passing through carbon monoxide, in the way that we have shown.

On the giant planets, C_2H_2 was formed by polymerization of CH, that is, by direct synthesis, and it was converted to methane by hydrogenation. In the presence of excess hydrogen, in a reducing medium, oxygen, carbon, and nitrogen form saturated hydrides, H_2O, CH_4 and NH_3, while on the terrestrial planets, where hydrogen is scarce but oxygen abundant, carbon dioxide and ammonia will be formed. Thus, it may be understood why Venus and Mars do not reveal the presence of an atmosphere of methane but of an abundant layer of carbon dioxide: 1 km of normal CO_2 on Venus and 4.40 m on Mars. On Venus, the oceans are iced and are unable to absorb this gas. On Mars, the absence of a hydrosphere prevents any absorption. On both, the absence of a biosphere prevents the formation of calcareous deposits.

In short, carbon exists on the giant planets in the form of hydrocarbons, whereas it exists on the terrestrial planets in the form of carbon dioxide.

The effect of the nascent water in the magma was not solely a conversion of the carbon of the metal carbides into carbon dioxide, but, generally, the metalloids were converted into oxygenated compounds. Thus, silicon sulfide, which Fremy believed to be abundant in the magma, gave silica and hydrogen sulfide:

$$SiS_2 + 2\, H_2O = SiO_2 + 2\, H_2S$$

The alkali earth silicides, decomposed by water vapor, gave hydrogen silicides, which were decomposed into silica and hydrogen:

$$SiMg_2 + 2\, H_2O = SiH_4 + 2\, MgO$$
$$SiH_4 + 2\, H_2O = SiO_2 + 4\, H_2$$

This is the origin of silica and of a part of the volcanic hydrogen.

Quartz probably crystallized below 400°C, which is the transition temperature of topaz. Finding inclusions of liquid CO_2, N_2, H_2, CH_4, and saline solutions of chlorides and alkaline carbonates (Renard, de la Vallée Poussin) on quartz proved that, at high pressure, CO_2 and an excess of hydrogenated gases were present in the magma and free oxygen was absent. According to Daubrée, diamond also sometimes shows inclusions of liquid CO_2.

The industrial synthesis of quartz is effected at 360–400°C in Na_2CO_3 solution and at 1000–2000 atmospheres pressure. Seeds are sown which grow at a rate of half a millimeter per day and which, in a few weeks, reach nearly 2 cm in length. They are transparent, without inclusions or flaws, and can be used in piezoelectricity.

The action of water on these nitrides produces ammonia:

$$FeN + 3\,H_2O = Fe(OH)_3 + NH_3$$

Gautier has found nitride of iron in the deep rocks, and Brun has found it in the lavas. Ammonia can also be produced by the action of water on the cyanamides formed from the occluded nitrogen:

$$CaC_2 + N_2 = C + CaCN_2$$
$$CaCN_2 + 3\,H_2O = CaCO_3 + 2\,NH_3$$

In the same way, the metallic phosphides will give phosphoretted hydrogen that can decompose into phosphoric acid:

$$2\,PH_3 + 5\,H_2O = P_2O_5 + 8\,H_2$$

which will be fixed on alkali earth bases to give apatite, P_2O_5CaO, which is found in almost all rocks in the proportion of some thousandths. It is the source of the phosphorus of the whole living world. Finally, the sulfides will give rise to hydrogen sulfide, sulfur, and sulfates such as kasterite, $CaSO_4$.

Celestial mechanics and the kinetic theory of gases show that, through the force of gravity, the earth was capable of retaining, at a high temperature (1500°C), its water vapor and, a *fortiori,* its carbon dioxide. But the same did not apply to the moon, which lost all these gases as soon as they appeared. The Moon has not lost its oceans: It has never possessed any.

Our satellite saw its physicochemical evolution arrested very quickly and, because of the lack of oceans and of an atmosphere, it has never known geological periods such as ours. It did not possess metamorphic rock, such as gneiss and granites, or any sedimentary terrain. It has been the same for Mercury, Mars, and Olbers protoplanet, all of which have seen their evolution arrested at this *lunar stage.* The stony meteorites may therefore indirectly give us data on the nature of the lunar rocks. Daubrée has shown that the stone of Juvinas was analagous to the lavas of Iceland, and was rich in pyroxene. The Igast meteorite of the seventeenth of May 1855 is analogous to a pumice stone.

We shall not therefore be in agreement with Gamow (16) who declared that the primitive oceans were fresh water, that the primitive crust of the earth was of granite, and that the moon has "carried away part of the granite."

In this primitive chemistry of the magma, this cosmic paleovulcanism, which occurred during the surface cooling, 4.5×10^9 years ago, we shall not consider the formation of our present-day rocks, such as gneiss, granites, and schistes, which are much more recent and the result of metamorphosis. The origin of the deposits has been for the most part biogenic. But the recognized geological periods only cover the last 500 million years of the globe's history.

This magmatic chemistry, which gave rise to water, carbon dioxide, and ammonia, has had very important physical effects on the surface of the earth and moon. These volatile substances have emulsified the superficial magma, and have given rise to enormous internal magmatic pressures, which have caused it to swell, providing these celestial bodies with their primitive relief, a relief preserved in its entirety by our satellite, as Puiseux (17) has authoritatively demonstrated.

The Primitive Marine Environment; The Volcanic Cycle of the Salt

The problem of the origin of life could not be dealt with without discussing the nature of the primitive marine environment. Because life is only able to originate in this environment, it is important to know its salinity and pH. A primitive ocean of fresh water would have been unsuited to the emergence of life.

Geochemistry shows that the alkalis and halogens are concentrated at the surface of the globe, the first in the form of alumino-sodium silicates, the second in the hydrosphere. The geochemical abundance is in accord with the abundance in meteorites and shows 20 times more sodium than chlorine. The marine salts contain mainly sodium (78% NaCl), whereas the oceans have a large excess of chlorine ($Na/Cl \sim 1/2$).

Since Boyle (1673), people have sought to explain the salinity of the ocean by the progressive erosion of the continental surface and the saline silting of rivers. The water of the rivers would in fact suffice to fill the ocean depths in as short a time as 40,000 years. But the best estimates of J. Murray led J. Joly, in 1902, to an age of the order of a hundred megayears for the oceans, some 40 times too short. In addition, the river salts differ greatly (80% carbonates, 7% chlorides) from the marine salts (0.2% carbonates, 70% chlorides). Some elements, such as boron, bromine, and iodine, which do not exist at the surface of the continents, could not have been brought to the marine environment by the river flow. Isolated basins, such as the Dead Sea, show very different salinities. A large amount of sodium which appears to escape from the erosion of granites is carried to the ocean. This is the "enigma of the salt," emphasized by P. Termier. One could not imagine a "salt cycle" being shut off by the atmosphere, in which the showers were local and the rainfall brought little sodium. The sodium of the upper atmosphere seems to be of cosmic (solar) origin.

Suess believed in the theory of the young volcanic water and the gradual formation of the oceans by the volcanic water. Behrend suggested that the oceans had acquired their salinity progressively through the volcanic exhalation. However, there is too much volcanic water for it to be young. The moon, despite its internal radioactive heat, produces neither vulcanism nor water vapor. On the other hand, the statistical volcanic saline exudate does not have the same composition as the marine salts.

If the oceans had condensed gradually, their salinity and their pH could not have preserved their delicate equilibrium. If, according to Rubey, the Na^+ ion concentration decreased by a hundredth, the pH would fall to 2.6, leading to the death of all marine creatures. On the whole geological and biological facts speak in favor of the perenniality of the oceans and of life.

Because the most favorable temperature for life is 44°C and the salt content of the plasma of higher organisms is from 8 to 9 g/l—whereas the salt content of the sea is about 33 g/l—Quinton (18) maintained that life appeared when the oceans were still warm (44°C) and low in salt (8–9 g/l). In fact, the organisms which have emerged from the sea have *carried the marine environment away with them* (E. Desguin). On Mt. Stephen (Canada), Walcott found annelids identical to present-day annelids, which indicated that the temperature and salinity of the Cambrian seas, 500 megayears ago, were identical with present-day conditions.

In 1938, we suggested that the alkali halides had, together with silica, constituted the primitive atmosphere of the globe, since they had formed a rapidly solidified sea (40 m) on the primitive vitreous crust. We showed (19) how the ocean waters were caused by the chemical evolution of the lithosphere during cosmic surface cooling. Thus, the condensed oceans acquired their present salinity. Vulcanism was regarded as a heat engine using the geothermal energy of radioactive origin, through the hypercritical marine water cycle. Volcanic water is not young and its abundance is such that all the water of the oceans would pass in the cycle in a few million years only. The carbon dioxide and nitrogen of volcanic origin are, on the other hand, young: It is they which nourish the atmosphere and the biosphere.

Van Nieuvenburg showed, in 1945, how the hypercritical water was capable of dissolving and transporting silica and salts, and even metals such as Cu, Ag, and Au. The water penetrating through the suboceanic faults assumes the hypercritical state at a depth of 13 km without ever boiling; the hydrostatic pressure (1670 atm) is always higher than the vapor pressure (225 atm). Carrying all its salts with it, the water *diffuses* through the subjacent rock just as hydrogen, helium, and neon diffuse through heated glass. It reaches the magma in hydrothermal pseudo-fusion at 850°C at about 50 km and brings about volcanic chemical activity. The alkalis are incorporated into the lava while the halogens form the acid fumaroles of the ocean vulcanism:

$$SiO_2 + H_2O + 2\ NaCl \rightarrow SiO_2{\cdot}Na_2O + 2\ HCl \qquad -Q$$

The high temperature fumaroles deposit the marine chlorides, NaCl, KCl, $MgCl_2$, and $CaCl_2$, and liberate free or combined halogens, NH_4Cl, $FeCl_2$, $MnCl_2$, Al_2Cl_6, $PbCl_2$, HF, and HCl. We may cite, for example, the Parace volcano in Colombia, the source of a river of hydrochloric acid, the Rio Vinagre. Fouqué showed that marine $MgCl_2$ was decomposed into HCl and MgO, which was incorporated into the lava. He found ten marine salts in the exhalations of Santorin. In this way, halogens and boron are restored to the marine environment where they accumulate. Potassium, much less abundant in the marine environment (K/Na = 380/10,560) than in rocks and sediments (K/Na = 27,000/10,560), seems to be precipitated in the form of glauconite.

The kaolin clays, the result of degranitization by carbon dioxide, form deposits buried deep in the geosynclinals. We have the reaction:

$$Na_2O{\cdot}Al_2O_3{\cdot}6\ SiO_2 + CO_2 \rightarrow Na_2CO_3 + Al_2O_3{\cdot}6\ SiO_2 \qquad +Q$$

At a depth of —13 km, the lithostatic pressure is some 2500 atm and the temperature reaches 374°C. The subjacent faults carry to them the alkaline fumaroles, born of magmatic thermal convection currents and formed of silica and alkalis which, transported by the hypercritical water, bring about their granitization, by crystallization of the quartz and feldspars. These are among others, the *fumaroles* of A. Michel-Levy (1893), the *filter columns* of P. Termier (1900), the *ichor* of Sederholm (1921), the *poren magma* of Eskola (1932), the *web* or *deposit* of Raguin (1932), and the *intergranular film* of Wegmann (1935). This granite will form the heart of the deposits that cover it, it will be carried down and folded by the tangential pressures engendered by oceanic vulcanism, and it will be exposed anew to atmospheric erosion. Thus, vulcanism, the formation of mountain chains and their erosion cause the alkalis to go through this closed cycle. Thermal activity brings sodium and borax to the surface again. The oceanic Na/Cl ratio is the result of an equilibrium between the rate of erosion of the granites and the volcanic activity. In this way, marine life has been able to subsist for close to four thousand million years and is halted at this "salt cycle" (20). It is possible to show how the primitive obsidian continents have been granitized at their bases by these currents of magma, which rise beneath the continents and sink under the oceans (convection whirlpools of H. Bénard, 1901), where they carry the heat which gives rise to the oceanic vulcanism. It follows from the theory that the sublimates of marine vulcanism (Santorin) are richer in halogens than those of continental vulcanism (Lake Kivu).

It is possible that the slow diffusion of the hypercritical water, that contains salts due to the vitreous rocks at high temperature, is accompanied by isotope segregations. Potassium would, because of this, be able to have different compositions in the Archaean and recent granites. The hypothesis therefore lends itself to experimental verification.

The Origin of the Atmosphere

It is probable that after the departure of the moon, the extreme centrifugal force to which the completely molten and incandescent earth was subjected had driven off the atmosphere which it might then have possessed. It was then that the rare heavy gases, A, Kr, and X, still retained by the protoplanet, were lost, so that they are considerably less abundant on earth than in the cosmos. The earth was practically devoid of atmosphere at this epoch. We shall show how the constituents of the present atmosphere have been produced—like the water of the oceans and carbon dioxide, which formed the calcareous deposits—from the chemical evolution of the lithosphere; we shall also describe the way in which they are bound in short-period cycles and how their abundance is the result of an equilibrium.

A number of authors have thought that the present atmosphere of the earth was primitive. Arrhenius believed that the primitive free oxygen was formed by the ultraviolet photolysis of CO_2 and that nitrogen was the result of the combustion of a primitive cyanogen in this oxygen:

$$C_2N_2 + 2\ O_2 \rightarrow 2\ CO_2 + N_2$$

These are the very two gases existing on Mars and Venus, but we have seen that cyanogen could not be compared to acetylene.

Tammann (21) attributed the free oxygen to the dissociation of water vapor when the earth was still incandescent; the hydrogen escaped. Goldschmidt (22) saw the origin of this gas in the dissociation of water vapor produced in the upper atmosphere by ultraviolet solar radiations shorter than 1850 Å. But in the spectrum of nocturnal luminescence, the OH bands only appear at altitudes above 70 km, where the water-vapor content is very low. Ozone, for example, which is formed from oxygen by such photochemical effects although produced at much lower altitude, at about 25 km, where the oxygen pressure is high, is present only in small amount; its total thickness corresponds to 3 mm of normal O_3 gas.

Harteck and Jensen (23) also attributed the atmospheric oxygen to this dissociation but Dobson, by accurately measuring the water vapor at high altitude, showed that its pressure was controlled by the low temperature (−83°C) of the equatorial tropopause (vapor pressure of ice: 1 barye), and that, since the altitude of the latter was only 17 km, no photolysis occurred there. Mars and Venus show no trace of oxygen or ozone.

It is a common and widespread error to conceive of the primitive atmosphere that existed after the condensation of the oceans as analogous to our present-day atmosphere, and to think that oxygen is only a residue, a gaseous excess, which had not reacted during the globe's formation. The attribution of this cosmic origin to the atmosphere is based on the fact that the galactic nebulae reveal the presence of oxygen and nitrogen. The error arises by considering the atmosphere similar to the hydrosphere and lithosphere. Now, although the hydrosphere and lithosphere are combustion residues from which it is impossible to extract any chemical energy, the free oxygen, on the other hand, represents a latent energy endowed with great chemical affinity. If this gas was not continually regenerated by chlorophyll action, it would soon be completely fixed by the lithosphere. It can persist in the free state only in the stars and nebulae.

So it is that lightning unites oxygen and nitrogen, in the form of NO and nitrates at one and the same time. Tropical rainstorms fix 400 kg of nitrogen per hectare per year. It is estimated that over the entire globe, lightning removes 10^{15} g of nitrogen annually from the atmosphere, that is, one millionth of the atmospheric nitrogen. At this rate, the nitrogen-oxygen atmosphere would disappear in only a million years.

But oxygen would disappear even more quickly. It is rapidly absorbed by the respiration of living creatures. High temperature lavas release combustible reducing gases, H_2, CH_4, H_2S, and S. They contain oxidizable ferrous oxides. The terrestrial crust contains, apart from the fossil carbon of biogenic origin, graphite and numerous sulfides, HgS, Cu_2S, $Cu_2S \cdot Fe_2S_3$, PbS, ZnS, and native metals, whose mass exceeds that of oxygen. It was Koene, who first showed, in 1865, how the oxygen of the atmosphere resulted from an equilibrium between its regeneration and fixation rates. Lord Kelvin, Arrhenius, T. L. Phipson, and W. Vernadsky reached the same conclusion.

It is clear that our atmosphere of oxygen is not of ancient origin but contemporary, and that it is entirely due to the chlorophyll function.

Oxygen is the gas characteristic of life from the cosmic point of view, and it is noteworthy that it does not appear on any planet, not even on Venus and Mars, in which the atmospheres contain more carbon dioxide than the earth's.

The chlorophyll function, a complex mechanism may, however, as we shall see, be described by the reaction:

$$Q + 6\,CO_2 + 5\,H_2O \rightleftarrows C_6(H_2O)_5 + 6\,O_2$$

Under the influence of a foreign energy Q, of cosmic origin—the solar light—the elements of carbon dioxide and water form carbohydrates and liberate free oxygen. The inverse reaction, we have seen, represents the energetics of the living world.

Nitrogen, absorbed continually by lightning and the nitrifying bacteria, is produced by vulcanism as free nitrogen N_2, NH_4Cl, and $(NH_4)_2CO_3$, but volcanoes release much less nitrogen than carbon dioxide. According to Rubey (24), the gases from Kilauea contain 23.5% CO_2 and 5.7% N_2. The gases of fumaroles and geysers contain 0.33% CO_2 and 0.05% of N_2, that is, 6.6 times more. This ratio should be compared with that of the mass of carbon dioxide contained in the sediments, 35×10^{21} g, to the mass of fossil atmospheric nitrogen, 4×10^{21} g, that is, $35/4 = 8.7$.

The nitrogen released by the lithosphere is accompanied by rare gases. It is, therefore, the young "cosmic nitrogen" of C. Moureu and A. Lepape, occluded in the original molten magma. It is estimated that the lithosphere contains 50 times more nitrogen than the atmosphere. Biogenic nitrogen, on the other hand, is devoid of rare gases. It is the "internal" atmosphere of the earth.

Argon-40 is a significant constituent of the present-day atmosphere (1%). It is not primitive but has been produced gradually, as Von Weizsäcker showed (25), by the disintegration of part of the potassium-40 of the lithosphere. It is not retained, either by the moon or by Mercury, but could form, as Brown suggested, part of the atmosphere of Mars.

This argon is also a product of vulcanism. Boato *et al.* (26) have tried, without success, because of the difficulties of the experiment, to identify an excess of argon-40 in gases sampled at Etna, Stromboli, and Vulcano. But they have succeeded with the gases from the Larderello soffioni. These gases—which accompany water vapor at high temperature (190°C) and pressure (14 atmospheres)—contain mainly CO_2 (90 to 99%), H_2, H_2S, CH_4, N_2, and the inert gases.

The ratio of argon-40 to argon-36, compared to that of air, has been found to vary from 1.1 to 2, for 19 samples from fumaroles and soffioni.

The analysis was carried out with a Nier mass spectrometer. The Larderello soffioni produce 4.4×10^6 g of argon-40 per year alone, while the total quantity of atmospheric argon-40 is 6.6×10^{19} g. If a thousand or so such sources existed at the surface of the globe, all the argon-40 could have been formed in this way since the terrestrial crust consolidated. This isotope must exist in the atmospheres of all the terrestrial planets and form part of the ionosphere of the moon.

References

1. H. Moissan, "Le four électrique." Paris (1897).
2. H. Douvillé, *Compt. Rend.* **159**, 221 (1914).
3. L. de Launay, "La science géologique." A. Colin, Paris (1905); "Traité de Métallogénie." Paris (1913).
4. A. Gautier, *Compt. Rend.* **132**, 938 (1901); **150**, 1564 (1910).
5. Melville Calvin, *Sci. News Letter* **76**, 359 (1959).
6. A. Dauvillier, *Compt. Rend.* **250**, 3925 (1960).
7. G. Chaudron, "Thèses." Masson, Paris (1921).
8. T. Graham, *Proc. Roy. Soc.* **15**, 502 (1867).
9. A. W. Wright, *Am. J. Sci.* **9**, 294 (1875); **12**, 165 (1876).
10. S. Meunier, *Compt. Rend.* **53**, 77 (1862).
11. A. Dauvillier, "Le Magnétisme des corps célestes" Vol. 2. "Variations et origine du géomagnetisme." Hermann, Paris (1954).
12. J. Pompecky, *Sitzber Preuss. Akad. Wiss. Physik.-Math. Kl.* p. 410 (1925).
13. O. Hahn, "Was lehrtuns die Radioactivität über die Geschichte der Erde?" Springer, Berlin (1926).
14. H. N. Russell, *Publ. Am. Astron. Soc.* **3**, 22 (1918).
15. A. I. Oparin, "Origin of Life." Dover, New York (1938, 1953).
16. G. Gamow, "Biography of the Earth." Viking Press, New York (1948).
17. P. Puiseux, "La Terre et la Lune" Gauthier-Villars, Paris (1908).
18. R. Quinton, "L'eau de mer, milieu organique" Masson, Paris (1912).
19. A. Dauvillier, "Genèse, nature et évolution des planètes." Hermann, Paris (1947).
20. A. Dauvillier, *Compt. Rend.* **242**, 47 (1956).
21. G. Tamman, *Zs. Physik. Chem.* **110**, 17 (1924).
22. V. M. Goldschmidt, *Fortschr. Mineral.* **17**, 112 (1933).
23. P. Harteck, and J. H. D. Jensen, *Z. Naturforsch.* **3a**, 592 (1948).
24. W. W. Rubey, *Bull. Geol. Soc. Am.* **62**, 1111–1147 (1951).
25. C. F. Von Weizsäcker, *Physik. Z.* **38**, 623 (1937).
26. G. Boato, G. Careri, and M. Santangelo, *Nuovo Cimento* **9**, 1 (1952).

V
THE PHOTOCHEMICAL SYNTHESIS OF ORGANIC MATTER

Organic compounds, being endothermic, could only have come into being at surface of the globe under the influence of a foreign, terrestrial, or cosmic energy, such as geothermal energy, atmospheric electricity, radioactivity, or high-quanta solar radiations.

We have seen that paleovulcanism was certainly capable of producing heterocyclic nuclei that were stable at high temperature, but the geothermal energy could not have given rise to the complex molecules that formed living matter, since they would have easily been destroyed by a rise in temperature.

Atmospheric electricity phenomena, such as lightning, were of great intensity during the condensation of the oceans, and were able to produce syntheses. In the present-day atmosphere, free oxygen and nitrogen are combined in the form of NO_2. This process (Birkeland-Eyde) has been utilized industrially. In the primitive humid atmosphere of nitrogen, lightning was able to produce ammonium nitrite according to Thenard's reaction:

$$N_2 + 2\,H_2O \rightleftarrows NH_4{\cdot}NO_2$$

In our view (1), this reaction is occurring at the present time in the atmosphere of Venus and is the cause of the yellowish aerosol mist covering the planet. We shall see the terrestrial importance of this reaction subsequently.

Electrolysis due to telluric currents or to electrochemical effects, such as the oxidation of metallic sulfides, could bring about the formation of traces of organic matter. It is known that the electrolysis of water between graphite electrodes produces mellic acid. Now, natural calcium and aluminum mellates, $C_{18}Al_2O_{12}\cdot18H_2O$, and iron and calcium oxalates are known. But a hypothesis like this is highly implausible. Mellic acid appears to have a biogenic origin and to be the result of the weathering of lignites.

Stoklasa (2) employed radioactivity and showed that its radiation could have brought about the synthesis of sugars. In fact, it is possible to calculate that, four thousand million years ago, radioactivity was about twice as intense as it is today, since the age of the principal radioelements are of the order of the age of the earth. But the radioactivity still remained extremely weak. On the most active granite terrains, it corresponds to the production, in normal air, of only a few pairs of ions per cubic centimeter per second, or of the same order of magnitude as the ionization caused by cosmic radiation at sea level. Moreover, these rays also have the power to destroy the organic compounds which they could form. Finally, they are abiotic. If the radioactivity had been much stronger, life would never have appeared at the surface of the globe.

All these energy sources of terrestrial origin are inadequate to produce, at the surface of the primitive marine waters, an abundance of ternary and quaternary organic substances endowed with rotatory power. The solar ultraviolet light alone is capable of this, but it is now known that it is absorbed by the screen of atmospheric oxygen. It is this element which, in all its forms—atomic oxygen in the ionosphere, molecular oxygen, and ozone—is the atmosphere's most effective absorbent. In the upper ionosphere and above an altitude of 150 km, atomic oxygen absorbs the solar radiation shorter than 910 Å, which brings about its ionization. Above 80 km, radiations shorter than 1850 Å are absorbed in producing photodissociation of molecular oxygen (system of Schumann-Runge bands). Finally, between 20 and 30 km, ozone absorbs the 2100–2950 Å band and limits the solar spectrum to this wavelength. The ozone is produced by oxygen-dissociating radiation shorter than 1850 Å and is destroyed by longer, 2100–2950 Å radiations, which it absorbs. Thus, a photochemical equilibrium is established, the result of which is a diffuse layer of almost invariant ozone totaling slightly more than 2.7 mm of normal gas (0.58 mg/cm^2). But the almost "metallic" ultraviolet absorption of this gas is so strong that below an altitude of 20 km the solar spectrum is, in practice, limited to 3000 Å. This limit depends on the zenith distance θ. At sea level, the following values are found.

$\theta°$	Å
0	2920
20	2970
53	3030
65	3080

Wigand (3) has obtained 2897.3 Å for the limit of the solar spectrum at sea level and 2896.0 in a balloon at an altitude of 9 km.

By means of prolonged exposures P. Götz succeeded, at Arosa (1800 m), in photographing the solar spectrum up to 2893 Å.

Fabry showed that, for the sun at zenith, 3 mm of normal ozone reduced the intensity of the 3000 Å radiation to a hundredth of its initial value, the 2950 Å radiation to a thousandth, the 2900 Å to a millionth, and the 2850 Å to 10^{-10}.

The energy of this solar ultraviolet, which is completely absorbed by the atmosphere, is a significant fraction of the visible radiation. To a considerable extent, the sun radiates as a black body at 6000°K, the temperature of the photosphere, and the maximum energy of the spectrum is in the visible region (5500 Å). For such a black body, the energy radiated in the continuous spectral region, at wavelengths below 3000 Å, is, theoretically, $\frac{1}{28}$ of the total radiation, and in the region below 2000 Å, $\frac{1}{500}$ of this radiation. The spectrum is limited to 1500 Å. A part of this radiation is stored in the ionosphere during the day and slowly restored during the night in the form of nocturnal luminescence. A remarkable experiment carried out at Los Alamos on the fourteenth of March 1956, at 2 A.M. local time, which consisted in projecting a charge of 9 kg of NO to an altitude of 85 km by means of a rocket, has proven the existence of atomic oxygen and has shown the magnitude of this energy. NO is capable of catalyzing the exothermal reaction $O + O \rightarrow O_2$. In the night sky, a luminous sphere appeared which had a diameter four times the apparent diameter of the moon and half of its brightness; the phosphorescence persisted for several hours, and slowly grew weaker.

Atmospheric ozone is very scarce at sea level (3 mg/100 m^3 of air) as well as above 55 km. The *lower atmosphere* is transparent to artificial ultraviolet up to 2200 Å. It was Hartley, in 1880, who discovered that ozone limited the solar spectrum; this was confirmed by the work of C. Fabry and H. Buisson. Thus, a narrow transparent band of atmosphere exists—a sort of "window"—that was predicted by Meyer (4) in 1903, between the absorption bands of molecular oxygen and ozone; many authors have looked for it at different altitudes. Lambert *et al.* (5) sought the 2020–2100 Å band at Vallot Observatory (4347 m) with a quartz optical double spectrograph, and avoided the diffused radiation in the 1900–2150 Å region. Schumann plates, exposed for 40 minutes, detected no radiation. Moreover, according to Bayeux (1919), the solar radiation does not produce any trace of ozone formation on Mont Blanc (4810 m).

Meyer *et al.* (6), connecting a photocounter to a spectrograph, thought that they had detected it on Jungfraujoch. At Pic du Midi (2757 m), we have looked for it (7) with a highly selective copper sulfide photocounter, that was sensitive solely in the 2100–2300 Å band, which enabled all optics to be avoided and it placed any diffused light out of reach. This radiation could not be detected. Its intensity would have been, at the most, of the order of that produced, in this region of the spectrum, by the flame of a candle 10 m away.

Vassy (8) calculated its probable intensity and showed that it could be observed only at high altitudes, unless the temporary existence of improbable "holes" of ozone is granted.

The existence of this radiation has been definitively established by recording high altitude solar spectra with the aid of rockets. On October 10, 1946, a V-2 launched in New Mexico (9) reached an altitude of 160 km. A series of spectra were recorded, using a parachutable concave grating spectrograph, ground on aluminum. At 25 km, the radiation, hitherto cut off at 2925 Å, reappeared, weakly, at 2200 Å. At 34 km, the spectrum extended to 2650 Å and reappeared between 2260 and 2100 Å. Finally, at 55 km, it extended continuously as far as 2100 Å, showing absorption bands between 2700 and 2900 Å. It was very intense. Measurements made on the continuous spectrum between 3200 and 2200 Å are given in the tabulation.

λ (Å)	Energy (μwatts cm^{-2} Å^{-1})	λ (Å)	Energy (μwatts cm^{-2} Å^{-1})
5000	10	2800	3
—	—	2600	1.7
3200	8	2400	0.7
3000	7	2200	0.3
		2000	0.0

The intensity at 2200 Å is therefore still $3/100$ of the visible radiation in the region of the maximum. Above the ozone layer, the spectrum is again limited, at 2000 Å, by the molecular oxygen.

On the twenty-first of February 1955, a rocket which rose to 120 km recorded the continuous solar spectrum up to 1550 Å. The spectra showed absorption bands up to 1680 Å and thirty chromospheric emission bands, belonging to nine elements, from 1817 to 977 Å, including the strong hydrogen band of the Lyman H α 1216 Å series and two magnesium lines.

If the ozone screen disappeared, this abiotic radiation would reach the surface of the sea and destroy all terrestrial life. Only marine life, protected by a few decimeters of water, would be able to survive. Fortunately, it is not possible artificially to destroy the ozone screen, temporarily and locally over a country, by sending a load of liquid ammonia to an altitude of 25 km.

The reaction between NH_3 and O_3 in the gaseous phase produces yellowish-white fumes of ammonium nitrite.

$$2\,NH_3 + 3\,O_3 = 3\,O_2 + NH_4NO_2 + H_2O$$

The oxygen is regenerated and immediately reconverted into ozone by the ultraviolet radiation.

In the absence of oxygen in the primitive atmosphere, the solar ultraviolet spectrum was still further limited in the ultraviolet, probably by ammonia. Initially, this gas mainly existed in a dissolved form and in a combined form as ammonium bicarbonate in the cooled marine waters. Today it is of biogenic origin. Volcanoes do not release ammonia but ammonium salts. The present atmospheric NH_3 level is variable. According to Schloesing (10), it averages 2 mg/100 m^3 of air and 0.2 to 7.2 mg per liter of rainwater. According to Duclaux and Jantet (11), the ammonia contained in the lower atmosphere would be sufficient to limit the solar spectrum to 2020 Å. Its absorption begins at 2265 Å and increases rapidly for the shortest wavelengths (12). Thus, for 2100 Å, a 1-cm thick layer of normal ammonia transmits only $\frac{1}{200}$ of the incident radiation.

Nitrogen is transparent up to 1450 Å (13). The rare gases do not absorb in this region of the spectrum. Carbon dioxide absorption begins at 1610 Å. Atmospheric carbon dioxide, the thickness of which is 2.4 m of normal gas, is completely transparent up to this point, beyond this point, it dissociates into CO + O.

Figure V-1 shows diagrammatically the position on the spectrum of the Schumann-Runge (O_2) and Hartley (O_3) absorption bands and shows the ultraviolet transparency band at 2200 Å. The first begins at

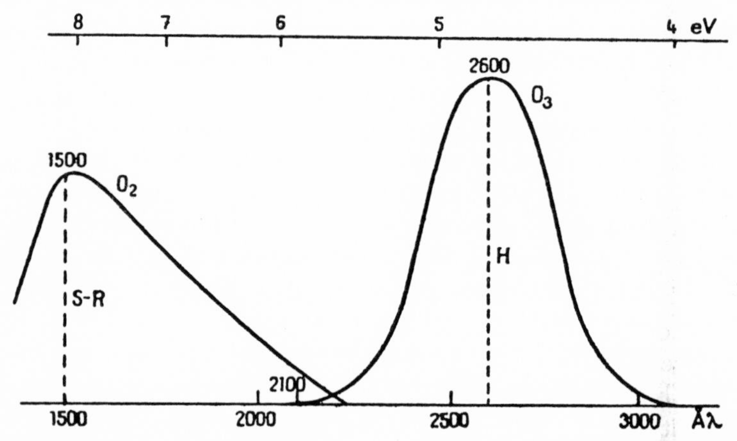

Fig. V-1. Diagrammatic representation of the ultraviolet absorption bands of molecular oxygen and ozone.

about 2200 Å and has a maximum at about 1500 Å. The second begins at about 3000 Å, has a maximum at about 2600 Å and terminates at about 2100 Å. It should be noted that the sterilizing abiotic ultraviolet radiation curve coincides to a considerable extent with Hartley's band. It begins at 3200 Å and has a maximum at 2650 Å. It is common to all cells because it is the absorption curve of thymonucleic acid, which is the sensitive element of the nucleus.

In second approximation, Fig. V-2 shows the absorption bands of

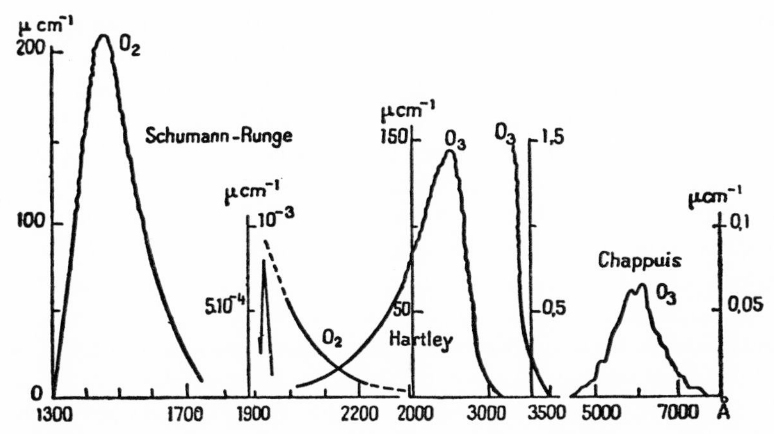

Fig. V-2. Absorption bands of ozone in the visible and ultraviolet spectrum. (After E. and A. Vassy.)

ozone in the visible and ultraviolet regions, and Fig. V-3 shows, in general form, the atmospheric transparency, on the logarithmic scale, from the Hertz waves to the solar X-rays, according to Siedentopf (14). The straight line R represents Rayleigh's diffusion, which varies as $1/\lambda^4$, and the straight lines T, the turbidity, which varies as $1/\lambda^a$. The optic "window" is indicated by F1 and the Hertz "window" by F2.

It is known that the near solar ultraviolet transmitted by the atmosphere is capable of producing chemical reactions: combination of chlorine with hydrogen, carbon monoxide, ethylene, etc. Chlorine acts on methane in the light to give the substituted derivatives, CH_3Cl, CH_2Cl_2, $CHCl_3$, and CCl_4. But the quantum is insufficient to dissociate the water and carbon dioxide. Berthelot and Gaudechon (15) showed that after a year's insolation no trace of formaldehyde was produced, but that the formaldehyde underwent photolysis, both in the concentrated form and in solution. As a matter of fact, H·CHO possesses absorption bands between 3500 and 2900 Å.

It has been known for a long time that the extreme ultraviolet radia-

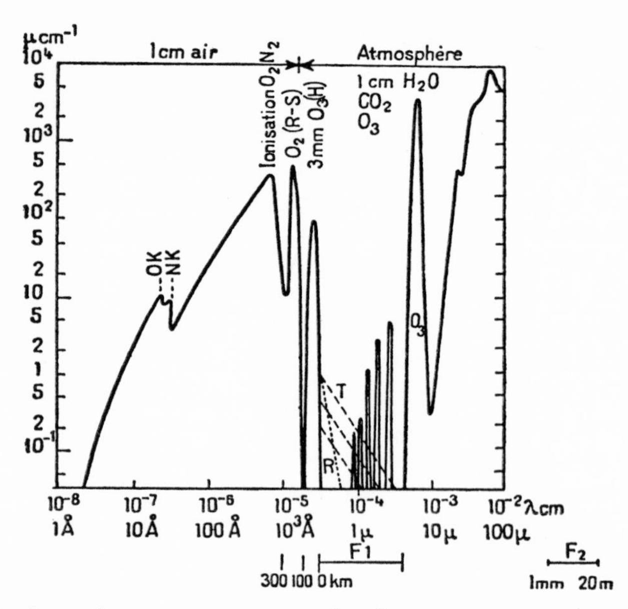

Fig. V-3. Atmospheric transparency in the electromagnetic radiation spectrum from the Hertz waves to X-rays (logarithmic scale). (After H. Siedentopf.)

tion is a powerful agent of organic synthesis. The spark and the arc are intense sources of ultraviolet producing ozone in the air, but Marcelin Berthelot's corona discharge tube is still more active, despite its low power, since it prevents any rise in temperature. Toward the end of the last century, Berthelot achieved a number of syntheses and studied the fixation of hydrogen, carbon monoxide, and nitrogen. The discharge tube was a high yield source of very far ultraviolet, but the chemistry of the electric discharge was not pure photochemistry because the gases were ionized. The discharge united nitrogen with O_2, H_2, H_2O, CS_2, and C_6H_6; these reactions could not be achieved by having the ultraviolet emanate from a quartz apparatus (limit 1800 Å). P. Thenard subjected a mixture of carbon monoxide and hydrogen to the ordinary electric discharge and obtained sugars. Similarly, in 1898, at the Solvay Institute in Brussels, H. Slosse synthesized sugars by subjecting a mixture of CO and H_2 in the Berthelot corona tube for 5 hours:

$$n\,CO + n\,H_2 = C_n(H_2O)_n$$

The liquid obtained reduced silver nitrate; it crystallized, and was able to ferment. Löb obtained formaldehyde by subjecting wet carbonic acid to the electric discharge:

$$CO_2 + H_2O \rightleftarrows H{\cdot}COH + O_2$$

It was from this aldehyde that E. Fischer achieved the synthesis of the sugars.

When the quartz mercury vapor arc lamp (R. Küch and T. Retschinsky, 1906) appeared, followed by the Kromayer and Heraeus lamps, pure ultraviolet photochemistry began to advance. A lamp, absorbing 600 watts, can, in fact, radiate 400 watts, of which 50 are in the yellow and green radiations and 40 in ultraviolet radiations. The infrared spectrum extends to 314 μ (Rubens) and the ultraviolet spectrum to 1800 Å.

M. Boll showed, with a Westinghouse lamp, that the intensity of the 2536 Å mercury resonance line increased as the square of the electrical energy expended (50 to 500 watts). With such a lamp, Herchfinkel (16) achieved the photolysis of carbon dioxide by an 80-hour irradiation.

Berthelot and Gaudechon (17), in a series of important works carried out with this apparatus, synthesized carbohydrates from carbon dioxide and water, and quaternary compounds from ammonia. The utilized spectral region was limited to the 2000–3000 Å range, and the effective wavelength was close to 2200 Å. In addition, they showed that radiations shorter than 2500 Å decomposed NH_3, PH_3, AsH_3, SiH_4, and $CoCl_2$, but not CH_4 and SiF_6.

These results have been confirmed by other authors. Stoklasa (18) has shown that the ultraviolet acted on a solution of $KHCO_3$ in the presence of nascent hydrogen to give formaldehyde, which condenses into sugars.

Taylor and Marshall (19) using the quartz mercury vapor lamp, have also fixed hydrogen on acetylene and carbon dioxide in sythesizing formaldehyde. The latter polymerizes rapidly giving rise to a number of complex compounds.

Groth (20), using the 1470 and 1295 Å xenon resonance lines, extended these reactions to the Schumann region and showed that a mixture of CO and H_2 gave formaldehyde, and a mixture of CO and CH_4, this same aldehyde, and glyoxal, CHO·CHO.

Very low yields were obtained in these experiments because the active radiation had to pass through two thin (2-mm) quartz walls and a few centimeters of air. A restricted solid angle was used, and the intense near ultraviolet radiations brought about the photolysis of the organic substances formed.

Bierry *et al.* (21) have reported that a large number of organic compounds possess an absorption spectrum in this region of the spectrum which brings about their photolysis. They have shown that their action caused extensive breakdown of a sugar, D-fructose, with the formation of formaldehyde and carbon monoxide.

Tian (22) showed that, in order to achieve effective syntheses with a minimum production of destructive antagonistic ultraviolet, it was necessary to utilize low-efficiency mercury arcs, that is to say, low-pressure, high-voltage lamps. He described a model lamp that was immersed in the experimental liquid medium. Today, for sterilizing liquids (E[ts] Gallois), alternating-current silica mercury lamps are built, and are lighted by a rare gas and hot oxide electrodes (length: 1 m, 500 volts, 50 milliamps), and are furnished with a pyrex soldered outer sleeve, in which the gases or liquids can be made to circulate. The continuous spectrum is very weak, the 2537 Å radiation represents 90% of the light, and the 1849 Å radiation, which produces ozone, is still detectable.

Figure V-4 compares the energy spectra of the high pressure (low voltage) mercury arc and the low pressure (high voltage) discharge.

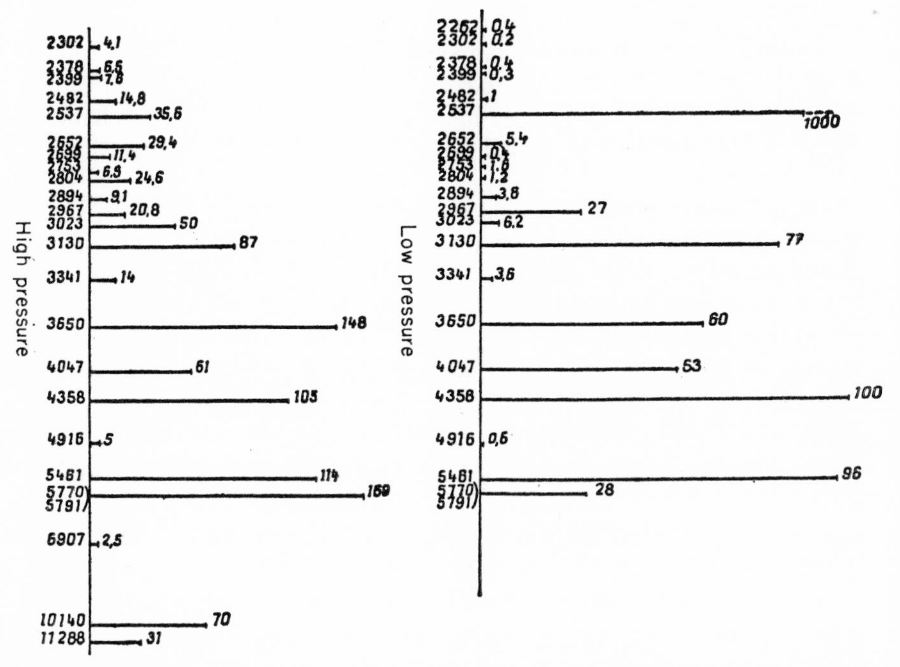

Fig. V-4. Energy spectra of the low pressure (high voltage) and high pressure (low voltage) mercury arc.

As other efficient sources, we may cite the hydrogen lamp, which gives a continuous spectrum extending up to 2200 Å, the iron arc, which gives numerous monochromatic radiations up to 2300 Å, and the high-frequency spark in water between aluminum electrodes (1900 Å).

Despite these improvements, the spectrum is limited to about 1800 Å, which makes the discharge tube much more efficacious than the mercury lamp. Comparative measurements have been made by Briner and Münzhuber (23) for the production of ozone. Their results are recorded in the tabulation.

Source	g O_3/kwh
Gaseous O_2 and 7-w mercury lamp	0.323
Gaseous O_2 and 28-w mercury lamp	0.169
Gaseous O_2 and 450-w mercury lamp	3.2
Gaseous O_2 and corona discharge tube	150
Liquid O_2 and corona discharge tube	300
Electrolysis of sulfuric acid	4 to 7
Maximum theoretical yield	1400

Since the original photosyntheses took place at the surface of the primitive marine waters, it is also important to know the transparency of water in this spectral region. Tsukamoto (24) showed that this liquid was still very transparent at 2000 Å, but it was completely opaque at 1800 Å at a thickness of 0.5 mm. In Schumann's ultraviolet (25), water vapor presents a continuous absorption band from 1880 Å to 1600 Å, the quantic energy of $H_2O \rightarrow (H_2O)^- \rightarrow H^- + OH$ dissociation is 6.6 ev. A photolysis occurs in the upper atmosphere above an altitude of 40 km. Sea water is less transparent than pure water below 2200 Å because of absorption due to the bromides which it contains. In the 4000–2500 Å region, the two main absorbents of sea water are NaCl and $MgCl_2$ (26). The salts in total, NaCl, $CaCl_2$, KCl, $MgCl_2$, $SrCl_2$, Na_2SO_4, $NaHCO_3$, NaBr, and H_3BO_3, introduce an absorption equal to ¾ of that of pure water. Photosyntheses are therefore evident only in a very superficial layer a few centimeters deep.

The synthesis of formaldehyde from CO_2 and H_2O leads to an oxygen release, equal in volume to that of the fixed carbon dioxide:

$$O = C = O + H \cdot OH \rightarrow O = C\begin{matrix} \diagup H \\ \diagdown H \end{matrix} + O_2$$

This aldehyde is, as Berthelot has shown, the starting point of a large number of carbohydrates. It has a strong tendency to polymerize and form glucides:

$$n(C \cdot H_2O) \rightarrow C_n(H_2O)_n$$

These glucides (glucose, saccharose, starch, and cellulose) are formed by the Würtz aldolization reaction to give alcohol-aldehydes:

$2\,(H{\cdot}CHO) = CH_2OH{\cdot}CHO$ (glycolic aldehyde)
$3\,(H{\cdot}CHO) = CH_2OH{\cdot}CHOH{\cdot}CHO$ (trioxymethylene)

$6\,(H{\cdot}CHO) = CH_2OH{\cdot}(CHOH)_4{\cdot}CHO$ (glucose)

From these compounds, polymerization is no longer able to occur by the simple addition of CH_2O molecules but it occurs by the regrouping of these compounds with the elimination of water, and the complexes obtained have the general formula $[C_6(H_2O)_5]_n$. In nature today, these compounds are manufactured directly by the chlorophyll of plants.

The glucose molecule $C_6H_{12}O_6$ can assume three structural forms: the linear carbonylic form, the furanic form (γ-oxide), and the pyrannic ring form (δ-oxide), which is stable and accounts for the synthesis of compounds higher by multiples of C_6; see Fig. V-5. We are therefore

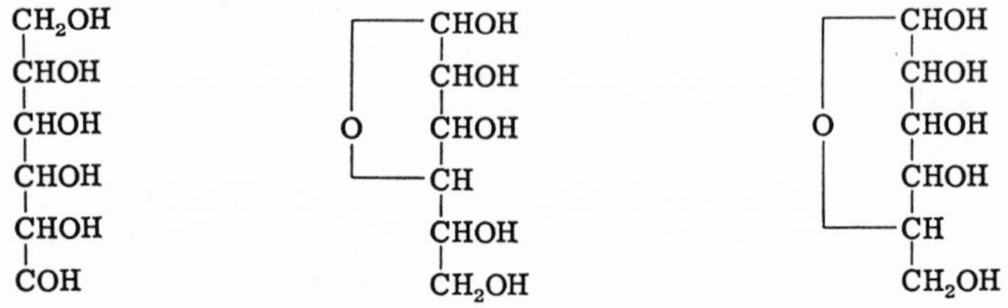

Carbonyl chain form | γ-Oxide furanose form | δ-Oxide pyranose form

Fig. V-5. Tautomerism of the glucose molecule $C_6H_{12}O_6$.

witnessing the appearance—this time in the cold and by the wet route—of the heterocyclic chains whose biochemical role we have already proclaimed. Two glucose rings, combining with the loss of a molecule of water, give rise to cellobiose (see Fig. V-6) which forms the elemental

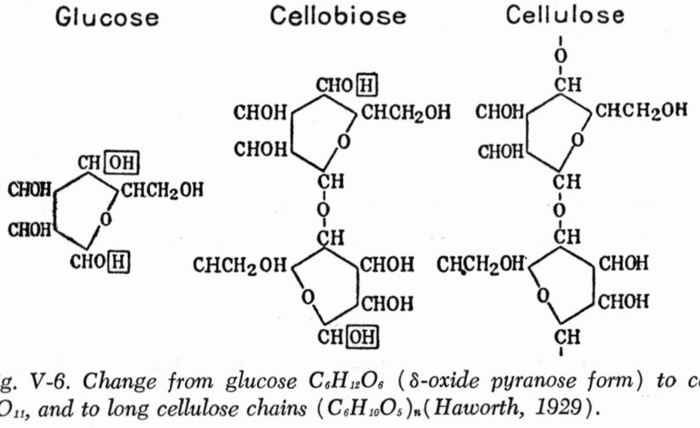

Fig. V-6. Change from glucose $C_6H_{12}O_6$ (δ-oxide pyranose form) to cellobiose, $C_{12}H_{22}O_{11}$, and to long cellulose chains $(C_6H_{10}O_5)_n$ (Haworth, 1929).

link of the long cellulose chains from which plant cells are built. The electron microscope showed the diameter of these elementary fibers to be some 10 mμ in the case of the ramie.

In the alginic acid of marine Algae, which also forms long polymerized chains, the primary alcohol CH_2OH is replaced by an acid radical CO_2H:

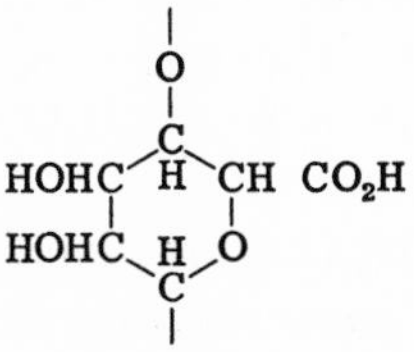

```
            |
            O
            |
            C
          /   \
    HOHC   H   CH  CO2H
       |       |
    HOHC   H   O
          \   /
            C
            |
```

But these ternary compounds could only have constituted food for the future living matter, for the latter is characterized by quaternary compounds in which nitrogen plays an essential role. Just as CO_2, dissociated into CO and O by the ultraviolet, can fix the hydrogen of water to form formic derivatives, so, in the same way, it can fix NH_3 to give the oily liquid *formamide*, as Berthelot and Gaudechon have shown experimentally:

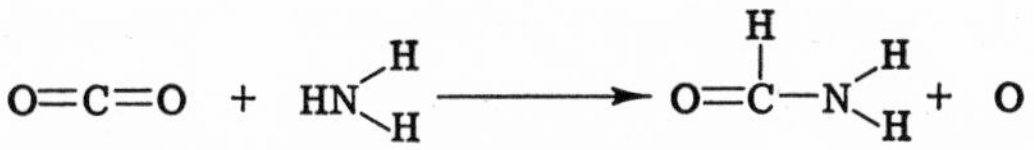

$$O{=}C{=}O + HN\begin{matrix}H\\H\end{matrix} \longrightarrow O{=}\overset{\displaystyle H}{\overset{|}{C}}{-}N\begin{matrix}H\\H\end{matrix} + O$$

Since this photosynthesis occurred at the same time as that of formaldehyde, under the same influence and in the same environment, the union of these two essential formic derivatives leads to the synthesis of glycine by condensation of a formaldehyde molecule with a formamide molecule:

$$H{\cdot}CHO + NH_2{\cdot}CHO \rightarrow NH_2{\cdot}CH_2{\cdot}COOH$$

It is a white water-soluble, sugar-flavored solid (Braconnot, 1820), that is produced by the action of sulfuric acid on gelatin.

Now glycine is the second term of a class of compounds, the amino acids, which are numerous and play an important role as constituent elements of living matter. By starting from glycine, E. Fischer was able to synthesize the polypeptides. Thus, the condensation of two molecules of glycine, with the elimination of a molecule of water, leads to glycylglycine:

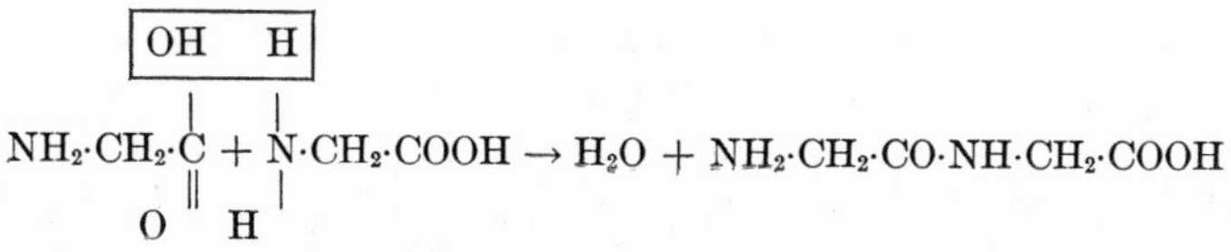

$$NH_2{\cdot}CH_2{\cdot}\underset{\underset{\displaystyle O}{\|}}{\overset{\overset{\boxed{OH}}{|}}{C}} + \underset{\underset{\displaystyle H}{|}}{\overset{\overset{\boxed{H}}{|}}{N}}{\cdot}CH_2{\cdot}COOH \rightarrow H_2O + NH_2{\cdot}CH_2{\cdot}CO{\cdot}NH{\cdot}CH_2{\cdot}COOH$$

and, in a general way, to the long chain amino acids: $NH_2 \cdot CH_2 \ldots CO{-}NH \ldots CH_2 \cdot COOH$ (CO: carbonyl; COOH: carboxyl). In 1903, Fischer succeeded in synthesizing a polypeptide of eighteen amino acids with a molecular weight of 1268.

Such chains, in which the molecules are bound by the carboxy-imide bond (the 4 biogenic elements):

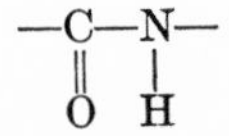

$$-\underset{\underset{O}{\|}}{C}-\underset{\underset{H}{|}}{N}-$$

have been demonstrated by X-ray diffraction. Thus, in silk fibers, according to Astbury, glycine and alanine alternate regularly:

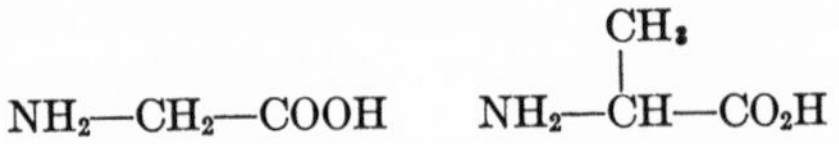

$$\underset{\text{glycine}}{NH_2{-}CH_2{-}COOH} \qquad \underset{\text{alanine}}{NH_2{-}\overset{\overset{CH_3}{|}}{C}H{-}CO_2H}$$

The same reaction, occurring in a molecule aminated in the δ- or γ-position, results in the formation of lactams (or the enol form, lactimes), which are heterocyclic compounds:

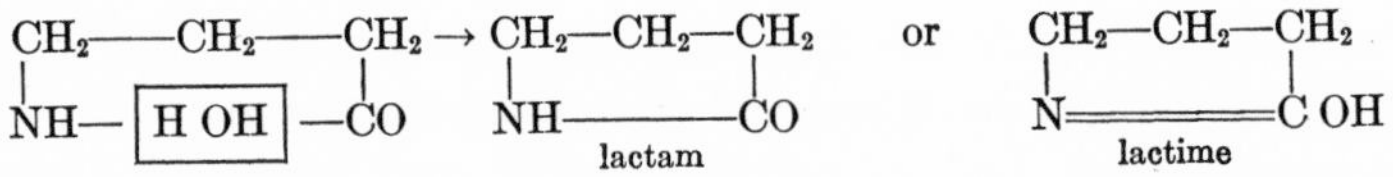

$$\begin{array}{l} CH_2{-}CH_2{-}CH_2 \\ | \qquad\qquad\quad | \\ NH{-}\boxed{H\ OH}{-}CO \end{array} \rightarrow \underset{\text{lactam}}{\begin{array}{l} CH_2{-}CH_2{-}CH_2 \\ | \qquad\qquad\quad | \\ NH{-}{-}{-}{-}{-}CO \end{array}} \quad \text{or} \quad \underset{\text{lactime}}{\begin{array}{l} CH_2{-}CH_2{-}CH_2 \\ | \qquad\qquad\quad | \\ N{=}{=}{=}{=}{=}C\ OH \end{array}}$$

With two molecules coupled head to tail, the diketopiperazine cycle is seen:

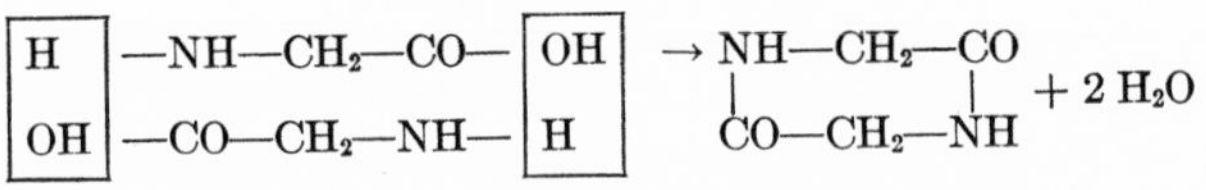

$$\begin{array}{|c|c|c|} \hline H & -NH{-}CH_2{-}CO{-} & OH \\ OH & -CO{-}CH_2{-}NH{-} & H \\ \hline \end{array} \rightarrow \begin{array}{l} NH{-}CH_2{-}CO \\ | \qquad\qquad\ | \\ CO{-}CH_2{-}NH \end{array} + 2\,H_2O$$

All these reactions are important in understanding how the required long chains (Wrinch's cyclol 6 theory) roll up so that molecules of very great mass remain largely spherical (The Svedberg).

Fischer's synthesis, which illuminates the pathway leading to first natural polypeptides, is not used today by living organisms. Proof of this is the fact that the carboxy-imide bond results in the disappearance of the nitrogen-containing group from the α position (it leaves only a single nitrogen-containing group at the terminal opposite that carrying the carboxyl). We know, however, that all natural amino acids carry a nitrogen in the α position. The terminal amine exists only in diamino acids (arginine, lysine), which are, in addition, animated in the α position.

Fischer's synthesis certainly explains the appearance of heterocycles but not those to which the most important anabolic (or synthesis) significance must be attributed. Chemists are agreed in separating the study of these compounds (anhydrides or imides of diacids, lactones, alcene oxides, etc.) from that of the heterocyclic compounds which are similar to the aromatic nucleus in regard to stability and properties. These (pyrimidine, pyrrole) are concerned in the constitution of the amino acids, to which the synthesis activity, for the building of protein molecules (histidine, proline, tryptophan), must be attributed. It is even probable that the acyclic amino acids (to which the origin of the albuminoids must be attributed) are never, at the present time, anything more than products of catabolism.

It is because of this characteristic of the true heterocycles that we have had to ascribe their formation to an epoch when a pyrogenic origin could be attributed to them. They may plausibly have had a catalytic role in the photosynthesis reactions of Berthelot and Gaudechon. This indeed is the best representation that might be made of the forerunner of the chlorophyll function, which, since it is not primitive, must have been preceded by much simpler processes.

It would be proper to turn again to Berthelot and Gaudechon's work using catalysts and high energy monochromatic radiation and chromatographic analysis of the very complex compounds obtained.

We do not exclude the fact that nitrogen itself, in the gaseous or dissolved state, may have entered into reaction and may have formed organic substances. Just as carbon dioxide and water combine under the influence of ultraviolet light, so nitrogen and water vapor may, combine directly to form ammonium nitrite:

$$N_2 + 2\,H_2O \rightleftarrows NH_4{\cdot}NO_2$$

This reaction is brought about by sparking (Thénard), that is to say, by lightning, or by the corona electric discharge (M. Berthelot), in other words, by the ultraviolet shorter than 1450 Å, which dissociates N_2. To an outside observer, our planets, at that time, looked as Venus does today. These yellowish white clouds, formed above the surface of the waters, rapidly dissolved. Now, according to Baly, Baudisch, and Heilbron, nitrites react on aldehydes in ultraviolet light to give formhydroxamic acid and oxygen:

$$NaNO_2 + H_2O + H{\cdot}CHO \rightarrow H{-}C\begin{matrix} \diagup\!\!\diagup NOH \\ \diagdown OH \end{matrix} + NaOH + O$$

The latter may be combined with $H{\cdot}CHO$ to give amino acids and pyrrole and pyridine compounds.

It is known that the lower atmosphere is very transparent today to the far ultraviolet as far as the Schumann-Runge absorption bands, and that flame is an intense source of these radiations. But the flames of the primitive volcanic eruptions could not have been utilized as a source of far ultraviolet, because, through lack of oxygen, they did not yet exist. No other type of flame is conceivable.

It could also be suggested that the terrestrial electric field, instead of having the present intensity of the order of a volt per centimeter, was 10^4 times more intense originally, and that the globe's surface was covered by an effluvium that caused Berthelot and Gaudechon's photosyntheses. But this field is maintained by the worlds lightning, and energy considerations are opposed to this hypothesis. More simply, it could be suggested that more powerful and more constant storms could have caused a permanent Saint Elmo's fire in certain regions of the globe. But we shall see that all these hypotheses are inadequate, and that resort must be made to certain crystals to explain the emergence of molecular dissymmetry.

S. H. Miller (27) showed, by chromatographic analysis, that it was possible to synthesize traces of glycine and alanine by causing an electric discharge to pass for 1 week in an atmosphere of the Oparin type, containing water vapor, methane, ammonia, and hydrogen. But we have seen that hydrogen and the hydrocarbons were excluded from the primitive atmosphere, that photosynthesis could not occur in the gaseous phase because of photolysis, and that these reactions could not account for the emergence of asymmetric molecules. In order to repeat Berthelot and Gaudechon's famous experiments correctly, ultraviolet photochemical reactions must be effected in an aqueous environment. The oxygen that was liberated by these reactions was retained in the dissolved state in water. A liter of sea water at 5°C contains 21.8 cm^3 of dissolved gas, that is, in volumes: $N_2 = 13.6$ cm^3, $O_2 = 7.3$ cm^3, and $CO_2 = 0.9$ cm^3.

The proportion of oxygen, 32%, is thus higher than in the air, 21%. If the oxygen had not been soluble in the water, marine life would not exist and would never have appeared on the earth. This solubility delayed the formation of an ozone layer above the waters' surface and the formation of an absorption band of oxygen limiting the solar spectrum to about 2000 Å. The organic substances escaped photolysis by the near ultraviolet by descending a few decimeters below the waters' surface.

Sea water contains about 3 g/l of magnesium and calcium sulfates and 0.005 mg/l of phosphates. The phosphates are more abundant in the cold plankton-rich waters of high latitudes. The marine environment, sugary and rich in proteins, therefore, contained all the elements necessary for the formation of the nucleotides, the bases of the nucleic acids characteristic of living matter.

Bernal (28) suggested that the clayey silts played an important role by adsorbing and concentrating these substances, and that the quartz microcrystals of the clays imposed their dissymmetry on them. However, quartz crystals always seem to be covered by an amorphous film of SiO_2, 100 Å thick. In fact, the almost hexagonal lattice of the silicates differs considerably from that of the polypeptides (10 Å instead of 4.8 Å).

Blum (29) suggested that graphite could have played a catalytic role in protein synthesis. Arcos and Arcos (30) emphasized the analogy that exists between the structures and periods of the reticular plane of graphite and the protein model.

Even if the ozone had always existed in the primitive atmosphere, the photosyntheses caused by the solar ultraviolet shorter than 2200 Å would have taken place. Thus, by a curious paradox, it is to the *abiotic* solar ultraviolet radiation that we turn for the creation of the organic matter which will become living matter. This radiation, which today stimulates the luminescence of the ionosphere and provides it with considerable warmth, was formerly used to produce photosyntheses at the surface of the marine waters and to maintain a primitive life there. (As a result of such considerations, the fine research of Hartley, Fabry, and Buisson becomes of capital geochemical importance.)

The Origin of Optical Activity

A large number of compounds are optically active. Some, such as quartz and calcite, are only so in the crystalline state. Their activity is due to their physical state and to their crystalline lattice that lacks an axis of symmetry (right quartz and left quartz with hemihedral facets). The optical dissymmetry of quartz is the result of the dissymmetric arrangement in the crystal lattice of molecules which are themselves symmetric. But a great number of organic compounds, such as oil of turpentine and camphor, are active in the dissolved, fused, or vapor state. Their activity is then due to their molecular structure. It is this molecular rotatory power, discovered by Biot in 1812 in a petroleum oil, which alone is of interest to us here. He showed that the molecule could not be superimposed on the image produced by a plane mirror, like the two hands, right and left, and proved the absence of certain elements of symmetry—center, plane, alternate plane of symmetry. This activity was considered characteristic of the biogenic origin of organic compounds for a long time. Activity is generally due to the asymmetric carbon atom. Le Bel showed that it could be due to the pentavalent nitrogen. However, the hexavalent platinum and cobalt of Werner's mineral—without carbon—complexes may also confer rotatory power. In 1914, Werner obtained the first mineral complex that did not contain carbon and did possess rotatory power:

$$\{Co[(OH)_2 \cdot Co(NH_3)_4]_3\}Br_6$$

the amino acids are active in the α position, and have at least one asymmetric carbon atom C*.

```
        H
        |
NH2—C*—COOH
        |
        R
```

The simplest imaginable optically active molecule would theoretically contain five atoms, but none is known that possesses less than eight. Le Bel and Van't Hoff have shown that this structure was essential for exercising a rotatory effect on the plane of polarization of light. Optically active compounds in the fused or dissolved state produce similarly active, pyroelectric and piezoelectric crystals (P. Curie). Born (31) has shown how wave mechanics could explain the existence or absence of a plane of symmetry in a molecule.

In 1860, Pasteur thought that optically active organic compounds could not be obtained without the intervention of life. Noting that all the immediate principles essential to life were dissymmetric, the distinguished scientist, in a report made in 1883 to the Chemical Society of Paris (32), expressed himself in these terms:

"When a ray of solar light comes to strike a green leaf and the carbon of the CO_2, the hydrogen of the H_2O, the nitrogen of the NH_3 and the oxygen of the CO_2 and H_2O form chemical compounds, and the plant grows, it is dissymmetric compounds which are formed.

"Life is dominated by dissymmetric activities whose enveloping, cosmic existence we are exhibiting. I have a feeling even that, primordially, all living species are, in their structure, in their outer forms, functions of cosmic dissymmetry. Life is the germ and the germ is life. Now, who could say what germs would *become* if the immediate principles, albumin, cellulose, etc. in these germs could have been replaced by their dissymmetric inverse principles? The solution would, on the one hand, consist in the discovery of spontaneous generation, insofar as that is in our power, and, on the other, in the formation of dissymmetric compounds using the elements C, H, N, S, and P if, in their movements, these atoms could be governed, at the moment of their combination, by dissymmetric effects.

"Should I insist on attempting dissymmetric combinations of simple compounds? I would cause these latter to react under the influence of magnets, solenoids, elliptical polarized light . . . finally, of all the dissymmetric activities that I could imagine.

"These dissymmetric effects, placed, perhaps, under cosmic influences —are they luminous, electrical, magnetic or calorific? Are they connected with the Earth's rotation or with the currents which produce its magnetism? It is still not even possible at the present time to hazard the slightest hypothesis on this matter."

However, in 1873, Jungfleisch showed that the molecular dissymmetry of certain organic compounds was not necessarily linked with the vital phenomenon. Starting from synthetic ethylene, he obtained resolved racemic tartaric acid, like that obtained from succinic acid of plant origin. He reproduced Pasteur's four forms of tartaric acid, but the hemihedral crystals had to be separated by hand, and an operator had to make the selection.

In 1886, A. Ladenburg achieved the brilliant synthesis of cicutine or conine, the alkaloid from hemlock: *levo*-α-propylpiperidine:

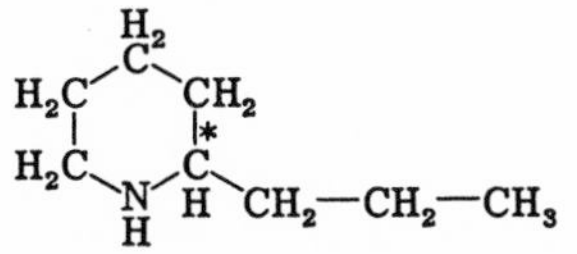

```
          H2
          C
       ╱     ╲
  H2C         CH2
   |           |*
  H2C         C
      ╲     ╱  ╲
        N   H   CH2—CH2—CH3
        H
```

In 1896, P. Walden discovered, on maleic acid, *optical inversion,* that is to say, the *complete* conversion of an optical isomer into its antimer.

Japp (33), in 1898, stressed the importance, from the vitalist point of view, of Pasteur's profound remarks. Chemical synthesis is incapable of producing an active compound to the exclusion of the antimer which balances it. In natural substances, on the other hand, active and not racemic compounds are encountered. In order to separate the *dextro*-rotatory form from the *levo*-rotatory form, the racemic compound must be combined with an active substance of natural origin, or cultivated moulds or the crystals must be sorted by hand, when they possess hemihedral faces—in short, in every case, resort must be made to a living organism. He concluded that the origin of the asymmetric compounds of a single species is as profound a mystery as the origin of life itself.

When the sorting of the enantiomorphs is carried out by microorganisms, one may speak effectively of the intervention of a living creature: *Aspergillus niger* consumes *levo*-tartaric acid, whereas the *dextro* acid is not satisfactory. On the other hand, *Penicillium glaucum,* grown on racemic *p*-tartaric acid, leaves a *levo*-tartaric acid residue. When the sorting is carried out by an observer selecting the hemihedral faces, it is no longer possible to speak of vital action because the choice could, it seems, be made as well by a photoelectric robot.

However, in Jungfleisch's experiments, it is no longer a case of *total asymmetric synthesis* being carried out under the influence of a dissymmetric physical agent and leading to the production of an active isomer in the pure state. Complete *rotatory power* has still not been reproduced artificially.

The work of Curie (34) on symmetry has shown that among the various physical agents which, theoretically, could influence chemical reactions—heat, rotation, acceleration, electric or magnetic fields, light, cathode rays—there is one only that possesses the required symmetry: circular polarized light. In 1894, Curie wrote: "I see dissymmetry everywhere. It is dissymmetry which creates the phenomenon." Let us not forget that Buridan's ass is the dead victim of symmetry.

Le Bel in 1874, and Van't Hoff (35) in 1894, suggested that this light would be able to bring about the photochemical synthesis of asymmetric compounds. Experiments tried by McKenzie on silver lactate were fruitless. In 1896, Cotton (36) discovered circular dichroism, that is to say, the unequal absorption offered in certain regions of the spectrum by alkaline solutions of copper tartrate, for circularly polarized right and left vibrations. He tried to make rotations of opposing signs appear in alkaline copper racemate solutions exposed to these radiations. Each circular polarizer was formed by a Nicol and a Fresnel parallelepiped. Six weeks exposure to sunlight brought advanced decomposition, but no rotation.

Byk and Mitchell (37) explained this result by showing that the *d*-tartrate, which shows a dichroism in the red, was only decomposed by the near ultraviolet, in which the absorption is the same for the right and left rotations.

Guye and Drouginine have sought unsuccessfully to demonstrate activity by carrying out chemical reactions in superimposed electrical and magnetic fields whose lines of force are parallel. So did J. Meyer, with the aid of a pencil of light, propagated parallel to the lines of force of a magnetic field of 180 gauss.

A number of attempts made by Freundler, Bredig, Cherbuliez, and Jeager also remained unfruitful. Positive results were meanwhile obtained by Kuhn and Braun (38) with the racemic form of ethyl α-bromopropionate, which showed the Cotton effect in the ultraviolet (2800 Å) and was decomposed by these radiations. The solution, irradiated by dextro circular light, was shown to be transiently dextro-rotatory (0.05°) with a 10-cm long polarimetric tube. A little later, Kuhn and Knopf obtained persistent 1° right and left rotations with the diethylamide of α-azido-propionic acid. Mitchell obtained symmetrical rotations with nitro-humulen in red light, which showed a maximum (0.3°) after 36 hours irradiation. Cotton (39) tried to make circular polarized light act on a mixture rendered inactive by compensation, and formed of two photochemically decomposable isomers that were endowed with circular dichroism.

Karagunis and Drikos (40) fixed chlorine on the triarylmethyl radical photochemically and, with CCl_4 as solvent, obtained a 0.08° rotation, which changed sign in the sense of the circular ray. The red solution, containing the radical, had absorption bands in the blue and yellow and the radiation used, circular polarized light, was selected in these regions of the spectrum.

Active molecules are derived from each other, like crystals or living creatures, and the problem of the origin of the first generative molecule is inseparable from that of the origin of life. That molecule would have been able to play the part of a dissymmetric catalyst determining the structure of all molecules formed subsequently.

Rectilinear polarized light exists in the solar light diffused by the sky and is produced in nature by reflexion, at an incidence of 37°, from the surface of water or on flat crystalline facets. The light from the sky is not polarized elliptically. Circular polarized light, which is obtained by causing rectilinear polarized light to fall on a quarter-wave plate—or a Fresnel parallelepiped, may also be produced naturally by polarized light falling on a birefringent quartz or spar crystal.

Becquerel, in 1880, noted that the terrestrial magnetic field could rotate the plane of polarization of rectilinear polarized light, and, subsequently, make the intensities of the *dextro* circular vibration and *levo* circular vibration unequal. The rotation of the plane of polarization is maximal in the magnetic meridian. It reaches 45′ to 5° on the horizon. Byk (41) sought the origin of natural asymmetry in a permanent, universal cosmic cause reigning over the whole earth. The reflection on water of the light from the sky, in part rectilinear polarized, would have given rise to elliptical, that is, partially circular, light. It is also necessary to appeal to the rotation of the plane of polarization of the rectilinear light by the geomagnetic field. It underlines the importance of Cotton's discovery which enabled the natural photochemical synthesis of dissymmetric compounds to be understood:

"Just as Wöhler's synthesis of urea constituted a powerful argument against vitalism, so the synthesis of a dissymmetric compound, to the exclusion of its enantiomorph, constitutes a new stage in the joining of the inorganic world with the organic world."

We shall not retain Vernadsky's hypothesis (42) in which the asymmetric factor resulted from the moon's separation from the earth. Not only does the rotation not have the desired character, but the formation of asymmetric macromolecules in the marine environment occurred much later, and for good reason.

Although an excess of one type of circular polarized light exists at the surface of the globe, in order to explain the appearance of natural asymmetric compounds, we shall rather refer, as suggested by Mathieu (43), to some fortuitous optical assemblage of a birefringent crystal and light polarized rectilinearly by vitreous reflection, which gave rise, by photochemical reaction, to an asymmetric macromolecule capable of self-reproduction. On the other hand, the production of an active species under the controlling influence of a substance which is itself active is known in the laboratory. Schwab *et al.* (44) quoted a case of dissymmetric catalyst in the dehydration and oxidation of fumes of racemic secondary butyl alcohol. Reduced metals (Ni, Pt), deposited in a very fine layer on quartz crystals of known optical sign, gave rise to an alcoholic residue that possessed rotatory power of the same sign as that of the quartz. The dissymmetric molecules produced today by plants are no longer produced by the influence of such photochemical reactions, but by the intermediate agency of molecules which are themselves dissymmetric, namely, the diastases.

The first photochemical asymmetric synthesis is, therefore, placed, on the *molecular scale,* at the edge of some briny shallow lagoon, surrounded by quartz or calcite crystals, at the foot of some tropical ocean volcano. Synthetic organic matter, having a density slightly above that of sea water (1.026 g/cm^3), would naturally concentrate at the bottom of these shallow lagoons. Calcite is transparent up to 2500 Å. These minerals had already existed for a long time. The enantiomorphous structure of quartz crystals indicates an environment which was itself enantiomorphous during the consolidation of the magmas. Cotton reported amethyst crystals formed of fine alternating strata of *dextro* and *levo* quartz. Spar is one of the very rare calcareous rocks of nonbiogenic origin.

Molecular asymmetry, the basis of life, would thus have been conditioned by the previous existence of the asymmetry of the crystal lattice, and this is certainly in harmony with geochemical evolution. Crystallization and life are two aspects of the organization of matter, of the appearance of order in molecular chaos, but although crystallization causes a static coordination to appear, life causes a dynamic coordination to appear (E. Desguin).

When the first self-reproducing dissymmetric macromolecule appeared, it imposed its structure on the complex organic environment as a whole by assimilating it, just as the crystal seed brought about suddenly the mass crystallization of the supersaturated solution into which it was introduced. This phenomenon could have occurred only once, and it is the result of a chance occurrence on the molecular scale. There is exactly the same probability of inverse symmetry appearing. Thus, on each planet peopled with living creatures, only one type of enantiomorph exists. If organization began with the appearance of pairs similar to the enantiomorphous creatures imagined by Pasteur, no vital competition would have opposed them. The nutrient value of their food would only have been reduced by half.

No more synthetic organic matter was produced subsequently, in a sufficient isolated amount and over a sufficient period of time, for there to be any chance of a new asymmetric synthesis occurring; and, because of this, living matter only offers one enantiomorphous form.

Asymmetric synthesis applies to the glucides as well as to the protids. These two groups of substances, intimately associated in the original ultraviolet photosynthesis, are still associated at the present time in chlorophyll photosynthesis.

The photochemical synthesis of the constituents of living matter is alone capable of effecting total asymmetric syntheses, and we find, in this fact, a decisive argument in favor of the photochemical theory of the origin of life.

Instead of resorting to foreign mineral crystals to produce polarized light, the latter can be supposed to have been formed in the organic substance itself, and in 1959, M. F. Bourdier suggested (private communication) that the origin of life could have derived from the properties of mesomorphic compounds.

Between the molecular chaos presented by the liquid state and the 3-dimensional lattice of crystal structure, various stages of progressive organization exist. Lehmann (45) discovered the mesomorphic compounds, which he inaptly named, "liquid crystals." Friedel (46) showed that these were not crystals, but that matter can present three discontinuous states: *amorphous, mesomorphous,* and *crystalline.* The mesomorphous state can itself have two forms: in the *nematic* form, the molecules are all oriented in the same direction and the substance has a fibrous structure. In the *smectite* form, they are randomly distributed in equidistant parallel planes, which makes them similar to the crystal lattice. This state, which is in tiers like that of soap bubbles, represents a stage of organization intermediate between the amorphous state and the crystalline structure. The molecules gradually lose their degrees of freedom. The mesomorphic compounds are anisotropic without possessing recticular structure. Thus, azoxyanisol fuses into a cloudy liquid at 116°C and acts on polarized light up to 134°C.

The filiform nematic and cholesteric compounds show cells and filaments in a state of division and multiplication. Friedel has shown that the cholesteric state is the form assumed by the nematic compound when it possesses enantiomorphic dissymmetry. Of all the mesomorphic compounds, the cholesterics are the only ones to be all of negative optical sign. Ammonium oleate and the ethyl ether of paraazoxycinnamic acid in

TABLE I

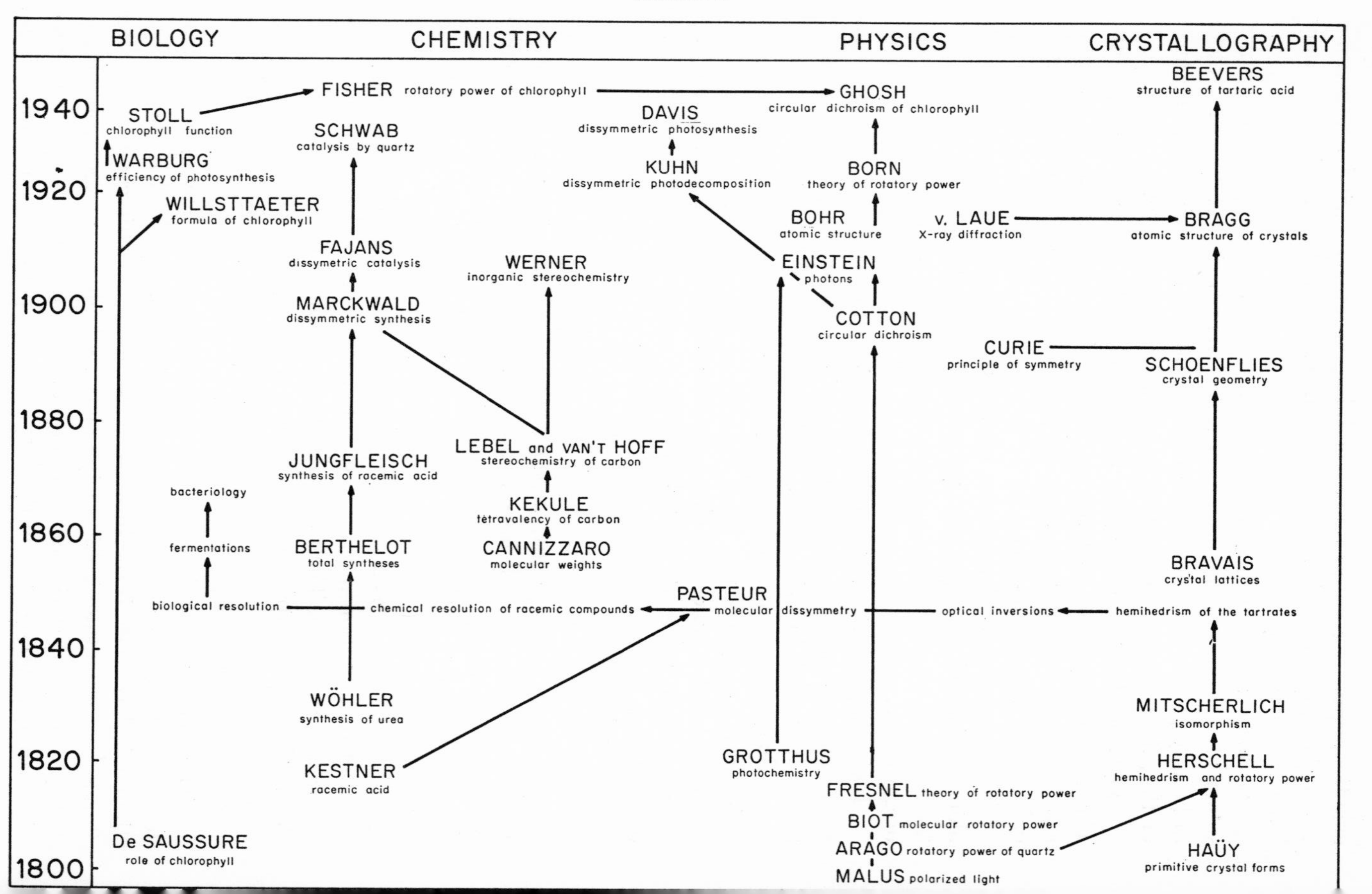
BIOLOGY
CHEMISTRY
PHYSICS
CRYSTALLOGRAPHY
1940
1920
1900
1880
1860
1840
1820
1800
STOLL
chlorophyll function
WARBURG
efficiency of photosynthesis
WILLSTTAETER
formula of chlorophyll
De SAUSSURE
role of chlorophyll
bacteriology
fermentations
biological resolution
FISHER rotatory power of chlorophyll
SCHWAB
catalysis by quartz
FAJANS
dissymetric catalysis
MARCKWALD
dissymmetric synthesis
WERNER
inorganic stereochemistry
LEBEL and VAN'T HOFF
stereochemistry of carbon
JUNGFLEISCH
synthesis of racemic acid
KEKULE
tetravalency of carbon
BERTHELOT
total syntheses
CANNIZZARO
molecular weights
chemical resolution of racemic compounds
WÖHLER
synthesis of urea
KESTNER
racemic acid
DAVIS
dissymmetric photosynthesis
KUHN
dissymmetric photodecomposition
PASTEUR
molecular dissymmetry
GHOSH
circular dichroism of chlorophyll
BORN
theory of rotatory power
BOHR
atomic structure
v. LAUE
X-ray diffraction
EINSTEIN
photons
COTTON
circular dichroism
CURIE
principle of symmetry
optical inversions
GROTTHUS
photochemistry
FRESNEL theory of rotatory power
BIOT molecular rotatory power
ARAGO rotatory power of quartz
MALUS polarized light
BEEVERS
structure of tartaric acid
BRAGG
atomic structure of crystals
SCHOENFLIES
crystal geometry
BRAVAIS
crystal lattices
hemihedrism of the tartrates
MITSCHERLICH
isomorphism
HERSCHELL
hemihedrism and rotatory power
HAÜY
primitive crystal forms

monobrominated naphthalene show bipartition phenomenon. In this latter case, spheres seem to be converted into rods or into long motile threads which divide, grow, and multiply. Although the rotatory power of quartz corresponds to 21° per millimeter, that of the mesomorphic compounds is enormous. Thus, paraazoxyanisol has a rotatory power of 5200°/mm, and is, therefore, very sensitive for a thickness of the order of a micron.

The structure of cholesteric and nematic compounds is not revealed by X-ray diffraction. Their planes are equidistant from 2000 to 80,000 Å and they, therefore, resemble the silver planes of Lippmann's interference plates. In the smectite compounds, the equidistances measure the lengths of the molecules and vary from 40 to 50 Å, as J. Perrin has demonstrated.

If such compounds could have originated in the primitive synthetic organic environment, they have directly engendered, in coacervates, circular light, the origin of the molecular dissymmetry of living matter. According to Bourdier, the solar light, on becoming polarized, could penetrate into each droplet and give rise to dissymmetric syntheses terminating in a self-reproducing structure. Experimental work on this important question should be undertaken.

Table I, taken from J-P. Mathieu (43), retraces the history of the discovery of optical activity.

References

1. A. Dauvillier, *Compt. Rend.* **243,** 1257 (1956).
2. J. Stoklasa, *Compt. Rend.* **156,** 646 (1913).
3. A. Wigand, *Physik. Zs.* **14,** 1144 (1912).
4. E. Meyer, *Ann. Physik.* **12,** 849 (1903).
5. P. Lambert, G. Déjardin, and D. Chalonge, *Compt. Rend.* **177,** 757 (1923).
6. E. Meyer, M. Shein, and B. Stoll, *Nature* **134,** 535 (1934).
7. A. Dauvillier, *J. Phys.* **3,** 29 (1942).
8. E. Vassy, Rev. *Opt.* **15,** 81 (1936).
9. W. A. Baum, F. S. Johnson, J. J. Oberley, C. C. Rockwood, C. V. Stain, and R. Tousey, *Phys. Rev.* **70,** 781 (1946).
10. T. Schloesing, *Compt. Rend.* **82,** 747, 969 (1876).
11. J. Duclaux and Jantet, *J. Phys.* **4,** 115 (1923).
12. Ferrieres, *Compt. Rend.* **178,** 202 (1924).
13. T. Lyman, *Phys. Rev.* **48,** 149 (1935).
14. H. Siedentopf, "Grundriss der Astrophysik." Stuttgart (1950).
15. D. Berthelot, and H. Gaudechon, *Compt. Rend.* **154,** 1803 (1912).
16. H. Herchfinkel, *Compt. Rend.* **149,** 395 (1909).
17. D. Berthelot and H. Gaudechon, *Compt. Rend.* **150,** 1169, 1327, 1517, 1690; **151,** 395, 478, 1349 (1910); **152,** 262, 376, 522; **154,** 1803 (1912); **156,** 1243, 1766 (1913); *J. Pharm. Chim.* (July 1910); *Rev. Gen. Sci.* **22,** 322 (1911). D. Berthelot, *J. Phys.* Jan. 1917.
18. J. Stoklasa, *Sitzber. Akad. Wiss. Wien Math.-Naturw. Kl. Abt.* **119** (1910).
19. H. S. Taylor and A. I. Marshall, *J. Phys. Chem.* **29,** 1140 (1925).

20. Groth: *Z. Physik. Chem.* **37,** 315 (1937).
21. H. Bierry, V. Henri, and A. Ranc, *Compt. Rend.* **151,** 316 (1910).
22. A. Tian, *Compt. Rend.* **156,** 1063 (1913).
23. E. Briner and A. Münzhuber, *Compt. Rend.* **242,** 1829 (1956).
24. Tsukamoto, *Rev. Opt.* **7,** 89 (1928).
25. A. Johannin-Gilles, *Compt. Rend.* **236,** 676 (1953).
26. J. Lenoble, *Compt. Rend.* **242,** p. 806 (1956).
27. S. L. Miller, *J. Am. Chem. Soc.* **77,** 2351 (1955).
28. J. D. Bernal, *Proc. Phys. Soc. (London)* **62,** 597 (1949).
29. H. F. Blum, "Time's Arrow and Evolution." Princeton Univ. Press, Princeton, New Jersey (1951).
30. M. Arcos, and J. C. Arcos, *Rev. Gen. Sci. Pures Appl. Bull. Assoc. Franc. Avance. Sci.* **63,** 89 (1956).
31. M. Born, "Elementare Quanten Mechanik." Springer, Berlin (1930).
32. L. Pasteur, *Rev. Sci.* **4,** 2–6 (1884).
33. F. R. Japp, *Rept. Brit. Assoc. Sect. Stereochemistry and Vitalism* (1898); *Nature* **58,** 452 (1898).
34. P. Curie, *J. Phys.* **3,** 393 (1894).
35. J. H. Van't Hoff, "Die Lagerung der Atome im Raume." (1894).
36. A. Cotton, *Ann. Chim. Phys.* **8,** p. 347 (1896); *J. Phys.,* **7,** p. 84 (1898); *J. Chim. Phys.,* **7,** 81 (1909).
37. S. Mitchell, "The Cotton Effect." Bell, London (1933).
38. W. Kuhn, and Braun, *Naturwissenschaften* **17,** 227 (1929); **18,** 183 (1930).
39. A. Cotton, *Ann. Physik.* **13,** 453 (1930).
40. G. Karagunis, and G. Drikos, *Z. Physik. Chem.* **26,** 428 (1934).
41. A. Byk, *Z. Physik. Chem.* **49,** 641 (1904).
42. W. Vernadsky, *Bull. Acad. Sci. U.R.S.S.,* No. 5, 633 (1931) (Russian).
43. J.-P. Mathieu, *Rev. Opt.* **9,** 353 (1930); "La synthese asymétrique." Hermann, Paris (1935); *Conf. Palais Découverte,* Ser. A, no. 161, 27 (1952).
44. G. M. Schwab, F. Rost, and L. Rudolph, *Kolloid-Z.* **68,** 157 (1934).
45. O. Lehmann, Flüsuge Kristalle Leipzig (1904); Die Neue Welt der Flüssiger Kristalle (1911).
46. G. Friedel, *Ann. Phys.* **18,** 274 (1922).

VI
THE ORGANIZATION OF LIVING MATTER. THE EVOLUTION OF LIVING CREATURES

In the previous chapters, we have shown how the primitive oceans were formed and immediately acquired their present salinity, how the latter, being the result of an equilibrium, remained unchanged owing to the volcanic cycle of the salt; how ternary, quaternary, and more complex organic substances originated at the surface of the marine waters by ultraviolet photochemical syntheses; how the molecular dissymmetry characteristic of life appeared in these molecules; and how, finally, a large amount of oxygen—present in the dissolved state in these waters—was liberated by these reactions. Thus, we have witnessed the genesis of a new type of geological formation, endowed with chemical energy and in a metastable state.

At what epoch of geochemical evolution can these reactions be placed? The geological periods, as a whole, cover hardly more than five hundred million years. One of the oldest fossils known, *Corycium enigmaticum* from Finland, a sort of graphitoid sack, was given the age of 1.6×10^9 years by K. Rankama in 1948, using C^{12} and C^{13} carbon isotopes. It would have been a pre-Huronian alga, that is to say, an organism already highly evolved. A. Holmes was able, through their radioactivity, to date ancient rocks containing traces of Algae, Protozoa, and Fungi spores. A granite from Lake Superior was 1.3×10^9 years old, and monazites from Southern Rhodesia have given figures of 2.6 and 3.3×10^9 years.

Sir G. H. Darwin, in his theory of the evolution of the earth-moon system, appealed to the powerful tides of the molten surface magma of the two celestial bodies in order to limit this evolution to 57 megayears. But in 1920, Milankovitch showed that the formation of a solid terrestrial crust of the order of a kilometer in thickness, which would have prevented any tide of this sort, had not required more than 10^4 years. The evolution of the earth-moon system should be attributed to the tides of the terrestrial hydrosphere, and it has not yet come to an end. The tides of the hydrosphere were enormous 4×10^9 years ago, the day lasted hardly more than 5 hours, and the temperature of the surface waters was much higher than that of the present-day tropical oceans (27°C). Photosynthesis could have begun when the temperature fell below 60°C, low enough to keep the volatile ammoniacal salts in solution.

It would therefore have been more than four thousand million years ago that organic matter appeared on the earth, and we have had all the time required—some two thousand million years—to reach the unicellular chlorophyllic alga.

This geological formation, sterile, hot, salty, ammoniacal, sugary, rich in proteins and nucleic acids, and in the presence of dissolved oxygen, was in the metastable state. Meanwhile, it would have been unable either to explode or burn, since it was in an aqueous environment. But it would have had a *tendency* to recombine with this oxygen and to be oxidized and release energy. Now, such a managed recombination of endothermal organic substances with oxygen, and liberation of energy, is exactly the characteristic feature of the energetics of life, whose *cause* and *necessity* we thus understand. It is possible to call the environment "living" although it was not yet accepted by living *creatures.*

In 1860, Pasteur discovered "life without air" or *anaerobic fermentation.* Leeuwenhoek had already recognized, in 1680, in yeast, a microorganism of a few microns that reproduced by budding, at the expense of its food. But fermentations could be produced by the purely chemical route by means of the juices extracted from moulds, that is to say, from numerous enzymes or diastases. In alcoholic fermentation, only a small part of the energy of the glucose, as measured by its heat of combustion, is utilized:

$$C_6H_{12}O_6 \rightarrow 2\ CH_3{\cdot}CH_2OH + 2\ CO_2 \qquad +22{,}000\ \text{cal.}$$

In lactic fermentation, there is no release of carbon dioxide:

$$C_6H_{12}O_6 \rightarrow 2\ CH_3{\cdot}CHOH{\cdot}CO_2H \qquad +22{,}500\ \text{cal.}$$

These anaerobic fermentations are inhibited by oxygen, as Pasteur showed. Instead, aeration brings about *respiration,* a more complete oxidation of the glucose, which can lead to oxalic acid:

$$2\ C_6H_{12}O_6 + 9\ O_2 \rightarrow 6\ CO_2H{\cdot}CO_2H + 6\ H_2O \qquad +2 \times 493\ \text{kcal.}$$

or to complete degradation:

$$C_6H_{12}O_6 + 6\ O_2 \rightarrow 6\ CO_2 + 6\ H_2O \qquad +674\ \text{kcal.}$$

In the case of fermentation, the microorganism develops only slightly, it grows rapidly by the respiratory process, which supplies more energy.

In every case, glucose is utilized in the form of phosphoric ester (Harden and Young, 1908), and phosphorus transport is effected by adenosine triphosphate (ATP). These two fermentations, passing through pyruvic acid, involve no less than fifteen diastases approximately.

In respiration, the carbon dioxide is derived from the decarboxylation of organic acids by diastase action. In this way, the pyruvic acid is decarboxylated by Neuberg's carboxylase (1911):

$$CH_3{\cdot}CO{\cdot}CO_2H \rightarrow CO_2 + CH_3{\cdot}COH.$$

Water is produced by the oxidation of hydrogen extracted and activated by enzymes such as Thunberg's dehydrogenases (1917). So, lactic acid may be dehydrogenated to pyruvic acid:

$$CH_3{\cdot}CHOH{\cdot}CO_2H \rightarrow H_2 + CH_3{\cdot}CO{\cdot}CO_2H.$$

Oxygen is activated by Warburg's iron-based red ferment (1924) and Keilin's cytochrome (1925), an electron carrier, which converts oxidized ferric iron Fe^{+++} into reduced ferrous iron Fe^{++}.

Complex cycles have been described by Szent-Gyorgii (1937) and Krebs. It has been shown (1) that in the Pasteur effect, the oxygen inhibits an enzyme fixing the phosphorus radical of ATP on fructose-6-phosphate. The thermodynamics of these reactions have been studied by R. Wurmser.

We can imagine that, in the absence of moulds, anaerobic fermentation was developed first of all in the primitive organic environment when ATP and some tens of enzyme species, which were not macromolecules but consisted of only about a hundred atoms of H, C, O, N, and P, appeared.

The Fungi, although heterotrophic, are nonmarine organisms and are too complex to be primitive and to have preceded the Bacteria.

The surface of the marine waters then showed the "tumultuous boiling of the vat," and released torrents of biogenic carbon dioxide. This indefinite living matter would have slightly resembled the plasmodium of the Myxomycetes, whose naked protoplasm contains countless nuclei and is analogous to an enormous amoeba. Its *mass* was the result of an equilibrium between its rate of synthesis by the ultraviolet radiation and its destruction by oxidation. We shall see later that the present biosphere, whose mass may be estimated at 2×10^{19} g, is renewed by the chlorophyll function in less than 50 years, granting a yield of the order of a thousandth. Knowing that the intensity of the active ultraviolet band (2200 Å) is 30 times weaker than that of the maximum of the visible spectrum, a quantity of organic matter of the order of the mass of the biosphere would be synthesized in a time as short as 50×30 or 1500 years, given a yield of the same order of size.

In this *living environment,* the first *macromolecule* endowed with *genetic continuity* would have been able to *reproduce,* as viruses and phages do today, within and at the expense of a living host.

When the complete synthesis of ATP and its enzymes is achieved in the laboratory, *the synthesis of life* will be achieved. If it is possible to synthesize nucleic acids, it will be possible to build new strains of virus endowed with genetic continuity, but the degree of complexity of organisms has always prevented their synthesis.

Sketches of organizations of molecular structures, are easily conceived in an environment whose complexity increases by itself with time. In this connection, Oparin emphasized the importance of the *coacervation,* or spontaneous separation, in dilute protein solutions, of drops of concentrated protein solutions or coacervates. This phenomenon occurred in sea water. We have seen that the amino acids form polarized chains with an alkaline and an acid end. They may be attached end to end, like magnets, with elimination of a molecule of water, and form long linear chains, or be juxtaposed, and form a filmy lattice whose two faces will have different catalytic properties.

Plateau in 1873, and Lord Rayleigh in 1890, produced thin layers of oils, paraffins, etc. as little as 2 mμ in depth on water. They were then studied by A. Pockels and, later, by H. Devaux. By means of simple, ingenious experiments, Devaux (2) showed, for example, that when stearic acid was spread on water, the hydrophyllic pole (the carboxyl) became immersed in the liquid. All plants are bounded by semiwettable layers by this method. The outer, nonwettable pole attracts molecules of hydrophobic, waxy substances and forms an impermeable cuticle. The simple contact of albumin with water forms coagulation membranes, so that no living organism can be wet by the surrounding environment. Devaux showed that these plasma membranes must cover all the small bodies contained in living cells. These membranes do not exceed 3 to 5 mμ in thickness. They function as double-faced catalysts, as molecule "traps." We see oriented molecular layers playing the role of Maxwell's demon in sorting and choosing molecules from the environment, as crystals do.

Folded in closed sacks, these membranes will act as double-faced catalysts; the inner poles will produce substances, such as fat, starch, and glycogen, and their outer faces will produce enzymes or soluble ferments, such as oxidases. A local chemical differentiation of the environment will be produced in this way.

The accumulation of matter inside such a closed membrane will cause it to swell, to the point where it may reach a size of the order of a micron. Self-reproducing macromolecules of deoxyribonucleic acid will play the part of genes comparable to the viruses of present-day nature. In this way, one can conceive of the outline of the first bacteria.

Boivin has shown that bacteria, just like animal and plant cells, possess a deoxyribonucleic acid nucleus in a ribonucleic acid-based cytoplasm. Bacteria possess a negative electric charge. They are destroyed by exposure to sunlight.

The heterotrophic bacteria certainly seem to be the ancestors of the living world: According to Vernadsky, their geochemical energy would be hundreds of times greater than that of the plant kingdom.

An ocean without bacteria would be uninhabitable for protozoa. They multiply rapidly in all terrains, even on the tops of mountains. They are found at an altitude of nearly 3000 m in the limestones and micaschists of the Pic du Midi, where they draw their nourishment from the nitrogen compounds carried by the snow and rain. Their prime function is to fix nitrogen. Without them, the plant kingdom would not exist. Some bacteria from thermal springs live at +88°C.

Bacteria are of the order of a micron (0.5 to 10) in size. Their mass is consequently of the order of 10^{-12} g. The bacterium, therefore, contains 10^{11} atoms, that is to say, as many atoms as the galaxy contains stars.

We have seen that the chemioautotrophic bacteria, on the contrary, could not be primitive since they are more complex than the heterotrophic bacteria. They effect more complex syntheses because of the multiplicity of their enzymes. Thus, *Nitrosomonas* synthesize all the amino acids, polyosids, nucleic acids, and proteins from CO_2 and NH_3 by means of specific enzyme catalysts. They play only a secondary geochemical role, and appear to us as late parasites utilizing the waste products of life, ammonia, hydrogen sulfide, and methane. The bacteria, whose oxidation of ferrous and manganese salts provides energy, are not primitive either, since they were never marine.

The bacteria, being the smallest living creatures, are also the most numerous, and are endowed with the greatest power of expansion, since their surface or exchange with the environment, and their ability to obtain food, is maximal. At optimum temperature and in the presence of abundant nourishment, some bacteria are capable of dividing indefinitely in less than ½ hour, their population doubles at this rate. The population increases by 2^n. After ten generations, that is, in 5 hours, the bacterium already has 2^{10}, that is, 10^3 descendants. Therefore, a balance must have been established, at the time they appeared, between their rate of expansion, the abundance of food, and their destruction by phages.

Haeckel believed that life was born at the bottom of the sea in the form of differentiated "plasma" of unnucleated cells (Monera). Also. when in 1857, during soundings made in the Atlantic for the laying of a submarine cable, Dayman brought a gelatinous substance resembling protoplasm from the great depths, Huxley thought that he recognized the original living substance and christened it "Bathybius Haeckeli" (1868). A closer examination showed that it was only a colloidal precipitate of $CaSO_4$ containing organic substances.

It is not at the dark and cold ocean bottom that the origin of life must be sought, or in the caverns, or in arid deserts, or in the atmosphere, or on the snows of polar and mountain regions, but only at the surface of the primitive warm, saline seas. All these places are unsuited to the emergence of life and were only peopled after the event. Emergent life has gradually extended its domain by colonizing these places by slow adaptation and has, finally, invaded the whole surface of the globe.

We have shown elsewhere (3) how the paleoclimatic problem could not be explained according to A. Wegener's ideas, or by a slow general drift of the rigid terrestrial crust—joined to the globe by the viscous bond of the molten magma underneath—in relation to the fixed axis of rotation. The polar migrations, which are necessitated by the paleoclimatic and paleomagnetic phenomena, must be explained according to Kelvin's ideas (1876), as the angular displacement of the spheroid (polhodie) with respect to the invariant axis of rotation. The energy is provided by the cyclic ocean currents, such as the Gulf Stream and Kuro Sivo, that is, ultimately by the solar radiation. It is restored in deep earthquakes, that is, in the end, in internal heat produced by the readjustment of the spheroid. Thus, the energy of the world's seismic activity is found numerically: It is 70^{26} ergs/year according to Gutenberg and Richter. If the Arctic, Antarctic, and mountain flora and fauna are so alike, despite the fact that they have always been separated by the insurmountable barrier of the tropical regions, it is because they have followed the migrations of the polar caps at the globe's surface, peopling, en route, the mountain regions in every latitude. It is through the agency of these islet samples, of these refuges, that they have acquired their uniformity.

The Appearance of Chlorophyll

This singular, primitive marine life, a tributary of the extreme ultraviolet, remained precarious. All organic molecules reaching the surface of the water underwent photolysis by the near ultraviolet. Every bacterium coming close to this radiation was killed by it. It was in these conditions that chlorophyll appeared.

Chlorophyll is a complex pigment, green in color, belonging to a whole group of pigments whose characteristic feature is the presence of a heavy atom; that of chlorophyll is magnesium. It is related to the respiratory pigment of the blood, hemoglobin, whose heavy atom is iron. The green chlorocruorine of annelids also possesses an iron atom. The blue hemocyanin of the blood of arthropods and mollusks (MW = 400,000 to 6,700,000) is copper based. The pigment of the chlorophylls, therefore, possesses an absorption spectrum in the visible region, and enables these radiations to be utilized for photosynthesis. The coloration of these molecules is due to the presence of pyrrole nuclei. Chlorophyllic photosynthesis again achieves the fundamental symbolic reaction:

$$Q + 6\,CO_2 + 6\,H_2O \rightarrow C_6(H_2O)_6 + 6\,O_2$$

The amount of energy Q is 674 kilocalories.

We have seen that the photolysis of water requires a high quantic energy, corresponding to the far ultraviolet. The $H_2O \rightarrow OH + H$ breakdown only takes place for a wavelength of 2400 Å. To effect this synthesis with a 6800 Å red light, it would be necessary to store up 4 quanta of energy. This spectral conversion is very important because the light energy is considerably greater in the visible spectrum of the sun; the maximum is in the yellow-green (5500 Å). It is completely analogous to the sensitization of silver halide photographic emulsion by colored pigments. These halides, sensitive only in the blue and ultraviolet, give, by the addition of dyes, ortho and panchromatic emulsions sensitive to the green, and then to the red. In 1874, Becquerel, and in 1905, Byk, sensitized photographic emulsions by adding chlorophyll. In the same way, Becquerel's photovoltaic effect (1839), which was manifested mainly in the blue for copper oxide, may be extended to the whole visible spectrum by dyeing with malachite green or methyl violet, as H. Rigollot showed in 1897 (see Fig. VI-1).

The works of Willstätter and Stoll established the existence of two chlorophylls of general formula:

(a) $C_{55}O_5H_{72}N_4Mg$
(b) $C_{55}O_6H_{70}N_4Mg$

Their percentage composition is as follows: C = 73.34, O = 9.54, H = 9.72, N = 5.68, and Mg = 0.34. These molecules, containing 137 atoms, are not macromolecules (MW = 904) and do not have genetic continuity. To build them, eleven genes are required to act in succession. If one of them is missing, the organism, although otherwise normal, remains colorless and dies after exhausting its reserves. The pigment is inactive *in vitro*, just as it is outside the chloroplasts, the small bodies which contain it. Study of the molecular structure of chlorophylls and

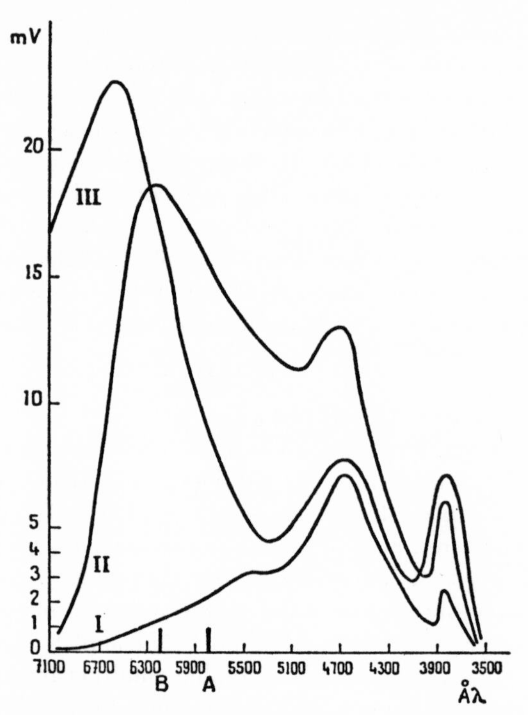

Fig. VI-1. Extension of voltaic effect of Cu_2O to the longer wavelengths by staining after H. Rigollot, 1897.

hemoglobin has shown them to be porphyrine-based, cyclic compounds with four pyrrole nuclei. Besides chlorophyll, two photosynthetic pigments with tetrapyrrole rings are known: the phycocyanine of some blue algae and the phycoerythrin of the red algae. The red sulfurous bacteria, containing bacterio-chlorophyll, produce photosynthesis as far as the infrared, that is, 9500 Å.

We have seen how heterocyclic molecules were first formed during the pyrogenic syntheses occurring at the time of the cosmic cooling of the lithosphere. One can see that, at the angles of the heterocyclic groups, aliphatic chains have sprouted like crystals of frost. These chains, joined by their ends, form adjacent aromatic rings (indole, carbazole, coumarone, flavone, phenazines, etc.), but, the free ends may also be attached to other cyclic rings. So it is that four pyrrole rings, C_4H_5N, alternating with four methine branches —CH=, form a cycle whose pyrrole nuclei are themselves groups. This is the structure of porphine that serves as a framework for chlorophyll as well as for hemoglobin. Figure VI-2 shows the structure of chlorophyll. The $C_{20}H_{39}$ radical is

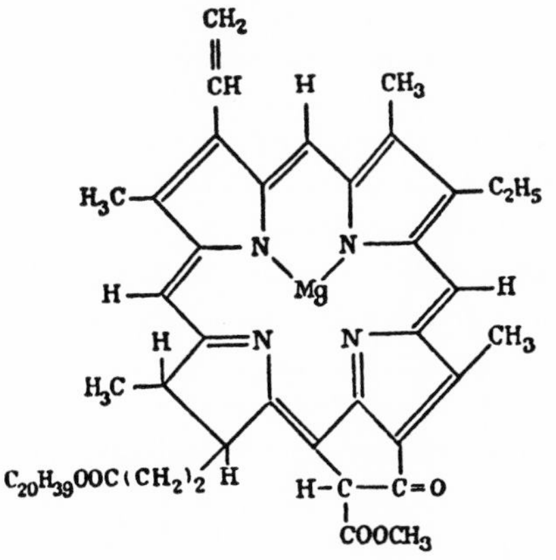

Fig. VI-2. Structure of chlorophyll.

that of an alcohol—phytol. Ghosh showed, in 1939, that chlorophyll in acetone solution possessed rotatory power and absorbed right and left circular light unequally in the red.

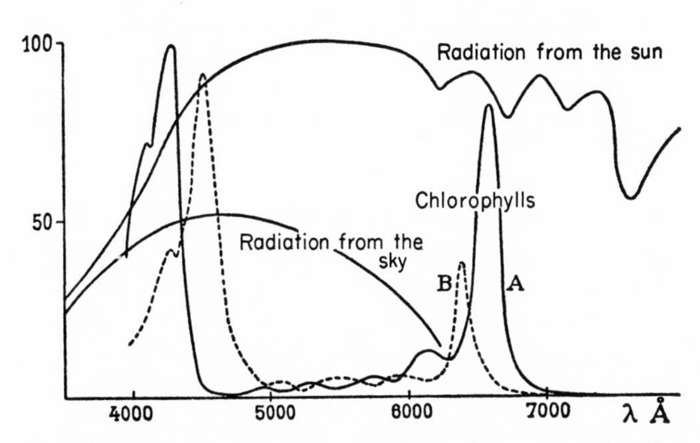

Fig. VI-3. Absorption spectra of chlorophyll a and b compared to the spectra of light from the sun and from the sky.

The absorption spectra of a and b chlorophylls are reproduced in Fig. VI-3 and compared with those of the sun and sky. The green color of chlorophyll is frequently masked by that of a colored carotenoid pigment whose absorption spectrum may vary considerably according to the spectral composition of the active radiation. In the marine environment, one observes, because of increasing depth, algae that are successively green, blue, brown, and red, according to the spectral variations of the transmitted light, which is soon reduced to a narrow band, 4300–4500 Å, at a depth of about a hundred meters. The algae with red phycoerythrin covering the chlorophyll absorb the blue green, and convert it into red. The pigment acts as a fluorescent screen and permits chlorophyllic assimilation to a depth of close to 100 m. The absorption of solar light in sea water is therefore not molecular but mainly biological. It is due to the pigments of the phytoplankton. The abundance of plankton and the presence of the red algae explain the various colorations of the seas.

Boussingault and M. Berthelot in 1864, and then von Bäyer in 1870, thought that photosynthesis took place through the intermediary of formaldehyde, but the work of Maquenne has shown that this was not so. Carbon monoxide, formaldehyde, and formic acid are poisons. Formaldehyde does not appear because the radiation of 6800 Å is used, and not of 2400 Å. Like the majority of photochemical reactions, it is accompanied by fluorescence that is in the infrared (7380 Å). This explains the "snow-clad" aspect of green landscapes photographed in the near infrared.

Wurmser (4) suggested that the reduction of carbon dioxide was a secondary effect due to the preliminary photolysis of water and this idea has been confirmed by the work carried out by Ruben using labeled oxygen and radioactive carbon, C^{11}, as tracers (20 mn period). The oxygen liberated after a few seconds came, not from CO_2, but from H_2O. This conclusion could also have been reached by simply noting that the analogous reaction:

$$CO_2 + 2\,H_2S \rightarrow H{\cdot}COH + H_2O + 2\,S$$

effectively liberates sulfur. The reaction is $H_2O \rightarrow H + OH$, then, $2\,OH \rightarrow H_2O_2 \rightarrow H_2O + O$, with release of oxygen. Atomic hydrogen combines with CO_2. In the light, chlorophyll is constantly synthesized and destroyed. Katz has shown that fluorescence is a measure of the consumption of energy by photosynthesis. *In vivo,* the tetrapyrrole nucleus would be formed from glutamic acid. The use of the artificial radioelements, C^{11} and C^{14}, the heavy isotopes, C^{13} and C^{18}, and of chromatography has allowed our knowledge in this field to advance rapidly. The photolysis of water is followed by reactions that can occur in the dark. Hydrogen and oxygen are transported by means of enzymes. After a few seconds 2-phosphoglyceric acid appears (Calvin and Benson, 1948):

$$\begin{array}{c} CH_2OH{-}CH{-}CO_2H \\ | \\ O \\ | \\ PO_3H_2\,. \end{array}$$

which leads to the lipids and triosephosphates, and finally to the glucides; saccharose appears after 30 seconds. This acid also gives pyruvic acid, $CH_3 \cdot CO \cdot COOH$, which, fixing H_2 and NH_3, leads to alanine and the proteins, which appear after 5 minutes. Twenty per cent of glucides are produced for 2% of proteins. Chlorophyll photosynthesis is the only process known in nature, at the present time, that reduces carbon dioxide with indirect liberation of oxygen. During its brief existence, the activated molecule effects this photolysis by absorbing some 10 quanta of red light. The energy absorbed by the coloring matter is transferred to the H_2O molecule.

In 1931, R. Audubert underlined the analogy between the photovoltaic effect and chlorophyll assimilation and showed that, in this effect, the light also shifted by photolysis of the water, the oxido-reduction equilibrium responsible for the voltaic photopotential.

Chlorophyll exists in the cytoplasm of bacteria and Cyanophyceae and, in the latter, it is masked by a blue pigment, phycocyanine. In plant cells, it is contained in the small bodies called *chloroplasts,* where it forms many flat follicles of 250 Å. The number of chloroplasts is of the order of 4×10^5 per square millimeter of leaf surface. In the chloroplasts, the pigmented lipid layers alternate with colorless protein layers. The pigments are oriented with their hydrophile poles in association with the proteins and their lipophile poles toward the lecithins (phosphated fats). The energy yield is very low, and does not reach 1%. On chromatographic analysis, the green leaf showed the following composition: chlorophyll a—$C_{55}H_{72}O_5N_4Mg = 62$, chlorophyll b —$C_{55}H_{70}O_6N_4Mg = 22$, xanthophyll—$C_{40}H_{56}O_2 = 9.3$, and carotene— $C_{40}H_{56} = 5.5$.

It is the yellow xanthophyll and red carotene that color the leaves in autumn after the chlorophyll is destroyed. The energy balance sheet of a green leaf may be summarized in the following way, as a percentage of the incident light energy: transmission = 10, absorption by evaporation = 49, reflection = 20, own radiation = 20, and utilization = 1.

The colored xanthophyll and carotene pigments transmit, to the chlorophylls, by fluorescence, the energy which they absorb.

Although chlorophyll synthesis requires the collaboration of pigments, chloroplast, and cytoplasm, it has been possible to effect particular syntheses *in vitro* using isolated chloroplasts. In 1939, Hill reduced ferric salts in the light:

$$2\,Fe^{+++} + H_2O \rightarrow 2\,Fe^{++} + 2\,H^+ + O$$

In 1951, Vishniac and his collaborators hydrogenated pyruvic acid into lactic acid and reduced 3-phosphoglyceric acid to fructose diphosphate, using suitable enzyme systems. A complete and up-to-date account of the chlorophyll problem will be found in Rabinowitch's work (5).

Fixed carbon dioxide and water, therefore, give mainly glucides, glucose, fructose, saccharose, and starch. The plant utilizes for its own requirements, hardly more than a sixth of the fats, vitamins, and alkaloids that it synthesizes. It thus allows the fungi and an abundant animal life to live at its expense.

Carles (6) showed that chlorophyll assimilation, which is very slight at 0°C, passed through a maximum at 37°C and fell to zero at 50°C. In exceptional cases, it was possible to find a reduced chlorophyll exchange down to —16°C for alpine plants, and down to —20°C for lichens.

The appearance of chlorophyll marked a decisive stage in the evolution of the living world. Living creatures, until then heterotrophic, became autotrophic and were able to produce abundant free oxygen on their own account. Large amounts of oxygen were liberated and the gas was released from the ocean waters, which gave rise to a layer of ozone in the lower atmosphere, and then, soon after, to an abundant atmosphere of molecular oxygen which henceforth prevented any photosynthesis by the extreme ultraviolet. Our present atmosphere had been created. This epoch coincides with the emergence of the first ferric oxide-based red rocks. The latter only appear at the end of the pre-Cambrian period, that is to say, more than 500 thousand million years ago, just before the recognized geological epochs. Limonite and hematite have been known since the Algonkian period. Therefore, more than three thousand million years of evolution would have been required to arrive at the Cyanophyceae. Life, protected by the ozone and oxygen, was henceforth able to invade the surfaces of the land masses, and respiration was able to become aerial.

The Cyanophyceae or blue-green algae are the most completely autotrophic organisms known. They achieve the most complex biochemical syntheses from CO_2, H_2O, and inorganic salts. They can fix nitrogen, like Winogradsky's bacteria. They live on the barest rocks, and were the first organisms to form colonies thereon. After the Krakatoa explosion in 1883, all trace of life had disappeared from what was left of the island. After a few years, the blue-green algae reappeared on the sterile lavas and ashes, forming, according to A. Ernst (1908), a dark green gelatinous layer. They withstand high temperatures and are found in Yugoslavia in the flora of hot springs at temperatures of +85°C. Those of Yellowstone Park withstand +61°C. Cuénot expressed the opinion that the existence of Cyanophyceae which withstand extreme temperatures indicated that they were first formed when the ocean surfaces were still at these temperatures. They withstand extreme desiccation and are often associated with lichens. They can live in highly saline environments, such as salt pans, and seem to be responsible for desert nitrate deposits. They have conquered the fresh waters and live equally well in acid environments (pH 4) and in alkaline environments (pH 9).

The history of the blood pigments could have been analogous to that of chlorophyll. As H. Munro Fox (7) wrote: "In the early geological periods, haemoglobin must have been synthesized in an animal for the first time. In the beginning, it could certainly have been a useless by-product of anaerobic life. Later, the usefulness of haemoglobin in certain cases as an oxygen carrier or oxygen reserve ensured its preservation through the laws of natural selection. In some cases, haemoglobin may still appear as an accessory product, without utility, and, in others, as the vestige of a substance which was useful in the past."

Unicellular Organisms

It would be foolhardy to insist on retracing the history of life to the appearance of the cell. In nature, at present, we are not acquainted with any transitional type between bacteria, Cyanophyceae, and flagellates, provided with a true nucleus, for only creatures with a silica skeleton or calcareous shell can be preserved in the paleontological archives. If the world of bacteria is already one of an unparalleled complexity, the cell is certainly more so. It was discovered by R. Hooke in 1661. It is characterized by the presence of tools, sense instruments or organs, that serve for locomotion, detection, and prehension. It contains a number of chemical substances: salts, sugars, fats, proteins, and bodies such as the mitochondria. But its main feature is the nucleus that consists of chromosomes formed by numerous genes, the source of its heredity. The evolution which led from the bacterium to the cell may have taken longer than the evolution which led from the protozoa to mammals, but it has left no trace in the paleontological archives. Only the diatoms, unicellular chlorophyllic brown algae, have left us their silica skeleton. They were very primitive and very numerous. *Euglena,* a unicellular flagellate with chlorophyllic plastids, presents us with a completed stage in the history of life for we find in it an autonomous, mobile, autotrophic being, reproducing by division, highly plastic, and endowed with psychism. To speak of "inferior" plants and animals is to fail to comprehend the perfection of the unicellular flagellate. The plants which derived from it, fixed and apparently devoid of psychism, represented a regression in evolution. The flagellates gave birth, at one and the same time, to the plant kingdom and to the animal kingdom that lived parasitically on it.

The algae, being more complex than the Cyanophyceae, are also more fragile, and none withstands more than 55°C.

Cellular partitioning seems to be a secondary phenomenon that is due to purely physical causes. The "artificial" cells with a copper ferrocyanide wall exciting the phenomenon of osmosis, studied by Pfeffer, Traube (1864), and Leduc (1907), are of little interest here. The cell walls are still plasma membranes possessing chemical activity and capable of secreting a calcareous or cellulose, or other type carapace. Among the Protista, the rhizopods possess pseudopodia capable of joining together, as rubber does, to form anastomoses. The cytoplasm of nucleated organisms contains small bodies, the *mitochondria,* of the size of bacteria, whose chemical activity is shown by the existence of circulation currents.

The mitochondria as a whole constitute the chondriome. Chloroplasts are chlorophyll carriers, leukoplasts contain starch, and blepharoplasts control vibratile cilia. Metabolic activity is localized in the hyaline part of the cytoplasm, the seat of oxidation phenomena and the part of the cell that is truly living.

The cytoplasmic currents are absent in the bacteria and Cyanophyceae, which do not possess mitochondria. The mitochondria resemble bacteria so much that Portier and Wallin suggested that they were symbiotic bacteria and that plastids were symbiotic algae. Cell life is due to the *controlled collaboration* of all these distinct bodies; the coordination devolved upon the nucleus. The cell possesses the complexity of a large automatic plant, all parts of which are closely interacting.

The nucleus of a 10-μ cell ($m = 10^{-9}$ g) does not exceed 1 μ, that is, 10^{-12} g. The genes (10^{-18} g) are included in the chromosomes. The *centrosome,* the body that controls the complicated phenomena of cell duplication (karyokinesis), does not exceed 1/700 of the chromosomes. The chromosomes contain thousands of genes. Study of the effect of radiation on the cell has supplied valuable information as to its structure. In 1923, F. Dessauer, with his theory of "hot points," put forward the quantic theory of this action and M. Blau, at the same period, calculated its probability. In the photochemical effects, the ratio $h\nu/KT$, the ratio of the quantum energy to the thermal agitation is very high and the photoelectron acts as a "hot point." The fine experiments of Holweck and Lacassagne (8), carried out with monochromatic ultraviolet (3000 Å) and X-rays (2.4 and 8 Å), whose photoelectron behavior in water was well known (see Fig. VI-4), have shown that the fatal effect of X-rays was like a veritable "target practice," and have enabled the dimension of the target to be determined. The apparent dose of radiation needed to cause a lesion—suppression of motility, of reproduction, deferred or immediate death—increases as the sensitive region of the cell body is reduced, as Fig. VI-5 shows.

In Fig. VI-5, the microorganisms are shown to scale. One protozoa, *Polytoma uvella* ($8 \times 12\ \mu$) one yeast, *Saccharomyces ellipsoideus* ($5 \times$

Nature of radiation	U.V.	X-rays		α-rays
Wavelength or course	3000 Å	8 Å	2 Å	3.9 cm air or 40 μ water
Diagram of the elemental of 10 quanta or of 10 particles d = 0.2 μ	No ionic trajectory	Trajectories of 0.04 μ with 35 ions	Trajectories of 1 μ with 174 ions	1000 ions or 0.2 μ (per α)
Absorption	molecular	atomic	atomic	atomic

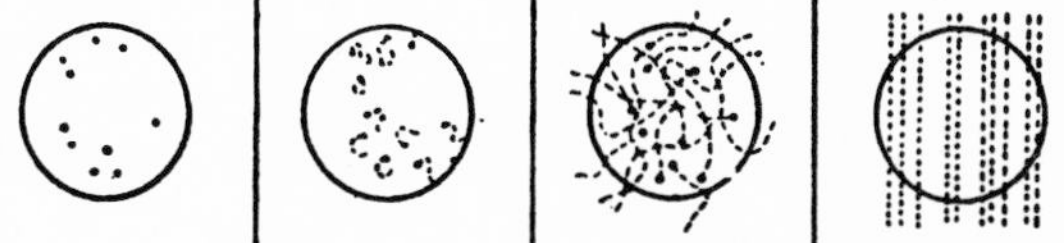

Fig. VI-4. Diagrammatic representation of the absorption of different radiations in a sphere of water, 0.2μ in diameter (after F. Holweck and A. Lacassagne).

7 μ), as well as two bacteria, *Escherichia coli* and *Bacillus pyocyaneus* are shown. It is seen that there is a remarkable correspondence between targets associated with immediate death (I.D.) and the centrosome, of those associated with deferred death (D.D.) and the chromatin of the nucleus, and of suppression of motility and the motor centers. To produce a lesion in a cell, the electron must pass through a cell body that encloses the sensitive zone. Since the electron's course is shorter, the longer the wavelength, the optimum yield is attained when the photoelectron's course

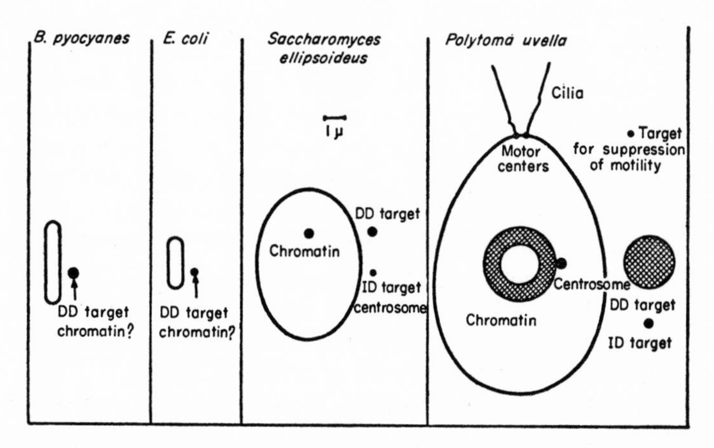

Fig. VI-5. Comparison between sizes of targets of deferred death (D.D.) and immediate death (I.D.), in different microorganisms and for cell bodies (after F. Holweck and A. Lacassagne).

is of the order of size of the sensitive zone. When the lesion is caused by a single quantum, the effect varies with dose according to an exponential law. When it is due to several quanta, it follows an exponential Poisson distribution. Thus, the shape of the "lesion curve" enables the size of the bodies to be calculated by a sort of statistical ultramicroscopy. The electron microscope has since confirmed these observations.

These experiments have been extended by Timofeeff (9), who investigated the probability of producing mutations by the action of X-rays. He found a target with a volume of $1/5 \times 10^{-6}$ cm^3, that is, a cube whose edge consists of ten atoms. If this cube is compared to the gene, the gene will therefore consist of a thousand atoms, which is the order of size of a macromolecule.

The Evolution of Living Creatures

Anaximander (611–547 B.C.) already knew that the first living organisms originated in the water of the seas, and that they had been gradually transformed into man. However, it was necessary to await the work of Buffon, J. B. Lamarck, and Darwin, and the advent of paleontology, for evolution to be taken as an accepted fact.

With the Protozoa, we are in the presence of free cell life. In the first stage of association, a *colony* of single celled organisms is being formed. The individuals give up a freedom which is too exposed for an autonomous collectivity possessing a single conscience. Although the *Volvox* is not primitive, it offers an image of such an association. It was discovered by Leeuwenhoek in 1700. It is a small (1 mm) creature, spherical, greenish, surrounded by a crown of vibratile cilia, and highly motile. It is an alga from the group of fresh water Chlorophycae. It comprises some 500 to 50,000 poorly differentiated cells. The isolated cell is viable but unable to reproduce. There is no apparent connection between the cells, but flagellar movements nevertheless suggest some coordination.

Among the multicellular creatures, we find, in contrast, differentiated cells. With hydra, we see fixed colonies of specialized individuals. With medusae, which are free individuals, sense organs appear such as tactile tentacles and rudimentary eyes. The annulated worms show bilateral symmetry and segmentation. Marine worms, of the genus *Convoluta,* are actually colored green by single-celled algae living symbiotically in their tissues. At a higher level, we see the emergence of colonies of multicellular specialized individuals, such as the social insects.

In Fig. VI-6 we have tried to represent the genealogical "tree" of the principal groups on the logarithmic time scale. Their width also shows, on a logarithmic scale, their geochemical importance, as defined by their mass, during the geological periods. But too many data are lacking to

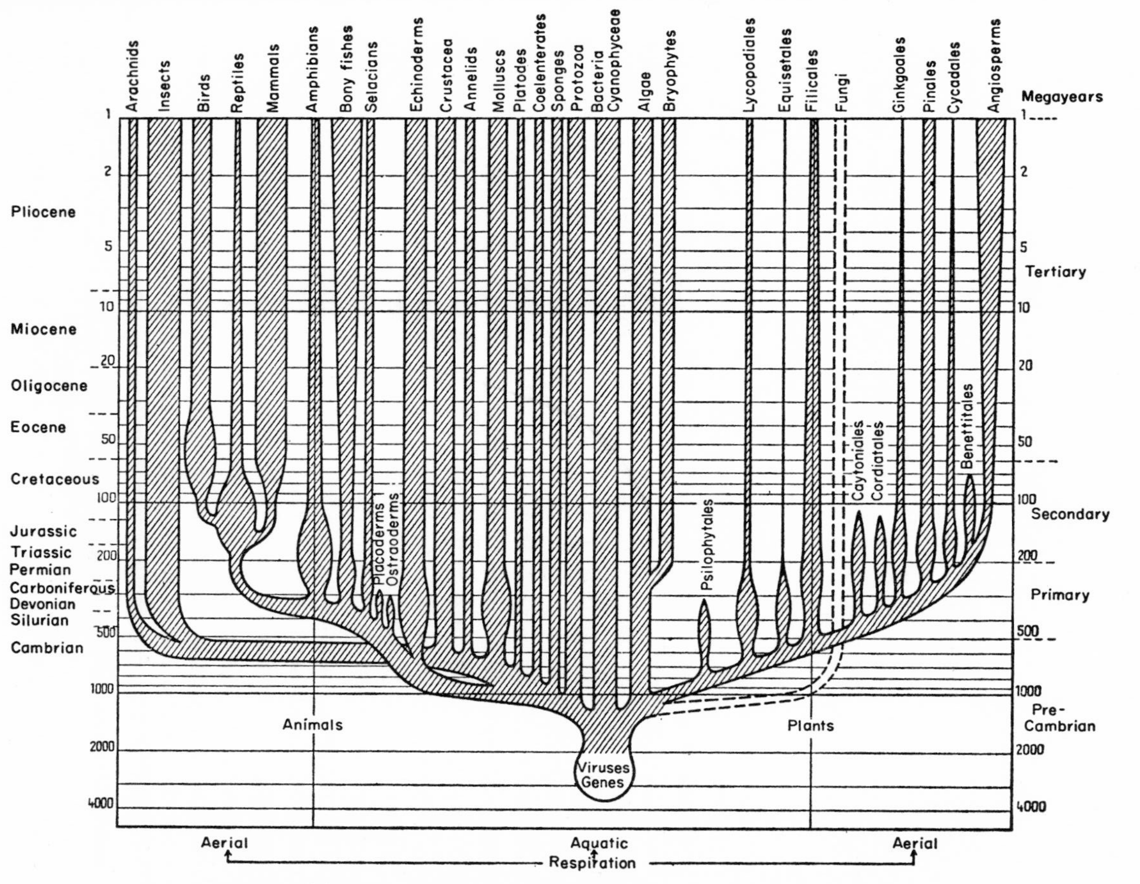

Fig. VI-6. Simplified genealogical "tree" of the living world, to logarithmic scale.

represent this quantitatively. It must be understood that the number of plants has always greatly exceeded that of the animals and fungi which live as parasites on them. If the graph were complete and quantitative, the total shaded areas would represent, on the logarithmic scale, the total volume of matter that has lived. Certain groups are extinct, others, formerly flourishing, are almost extinct today. Only secondary importance must be attached to the mode of respiration, aquatic or aerial. Yet, nothing demonstrates better that life was born in water. It is also seen that the lowest and most primitive forms still persist at the surface of the globe.

Evolution results, on the one hand, in the social insects, and on the other, in the birds and social mammals, all breathing air and all having highly developed psychism.

The fungi, which have no chlorophyll and are parasites of plants, manufacture glycogen instead of starch and possess a chitinous membrane. They may be carnivorous like nematode catchers. They have animal features and are difficult to explain, but their chemical nature places them close to the plants. Thus, the *Lactarium* secrete polyisoprenes identical to those of rubber. It is seen that invertebrate evolution had already come to an end in the Cambrian period and that vertebrates evolution began in the Silurian period with the fishes. In fact, the most ancient vertebrates, the cephalospid fish, date from 350 megayears ago. Amphibians already possessed facial expression, voice, and hands. The extinct primary plants formed coal. Plant evolution ended 60 megayears ago. Animal evolution is still continuing, since man appeared less than a million years ago. The letter S indicates the appearance of the monkeys in the Miocene period. In the face of this venerable antiquity of life, one should no longer be surprised at its infinite complexity. For thousands of millions of years, millions of species, each one made up of many thousands of millions of individuals, have evolved, and everything that was possible has had a chance of being realized. It is also seen that time does not play an essential role in evolution. The fishes are much older than the birds but remain much less highly evolved. They are only perfectly adapted to their environment. In 1908, Noel Bernard proposed an infective theory of evolution, according to which, the virus would be incorporated into the chromosomes as a new gene, suddenly altering the heredity.

Fritz Muller (1864) showed how the embryogeny of living creatures was only an abridged repetition of their genealogy. "Ontogeny is a short recapitulation of phylogeny" (Haeckel, 1866).

To A. Weismann (1902) are due the notions of *soma* and of *germen* (germ plasm), which cannot be reduced to each other. Cuvier's principle of the conservation of the individual form through continuous material changes, is attributed to the germen. But the experiments of P. Brien have shown that this division is too absolute. Besides the germen of the sexual cells, which are continually replaced and which, because of sexual reproduction, are immortal, and of the soma of the body cells, a third type of tissue must be considered, the neurons, that form the nerve cells, which are indivisible and irreplaceable.

Observation and experiment have confirmed that acquired characters are not inherited. However, the experiments only bear on a small number of generations and their length is insignificant compared to that of evolution, so that the extrapolation is not conclusive evidence. A species is represented by thousands of millions of individuals who pass through millions of generations each million years for nearly a thousand million years. There are not only nuclear but also cytoplasmic genes: these are plasmagenes. Bacteria may acquire, through heredity, resistance to the action of a poison; this can be explained by chemical modifications persisting in the descendants of the modified organisms. The bee, by artificially treating its embryos, practices ectogenesis and is able to determine the formation of a Queen, worker, or soldier. Instinct, which sometimes appears as a hereditary habit acquired through psychism, does not support Weismann's conception.

If evolution is not due to the influence of the environment, as Lamarck insisted, what then is its cause? In 1901, H. de Vries demonstrated the sudden, spontaneous appearance of new characters, immediately hereditary, in plants, and showed that the laws of hybridation, formulated by G. Mendel in 1865, were true. Naudin observed, in 1875, that even when a very considerable change occurred, it arose suddenly in passing from one generation to the next. The genes are therefore not immutable, and they may be modified spontaneously by these strong or minimal *mutations,* and may give rise to new characters. It seems that evolution has followed a similar, parallel course in the isolated regions of the globe. Thus, the marsupials of Australia have differentiated—independently—like the mammals of the other continents.

Weismann's theory and the discovery of mutations by de Vries were supported by a physical explanation when the biological effect of radiation was discovered. The very first X-ray irradiation experiments, such as those of Gilman and Baetjen, in 1904, on *Ambystoma* eggs, immediately showed the production of monsters. In 1927, J. H. Müller established that X-ray irradiation of *Drosophila* enabled the natural mutation rate to be increased up to 150 times. *Teratology,* or the artificial production of monsters, became, in the hands of E. Wolff, a remarkable technique in which X-ray doses were able to be localized on tiny, well-defined regions of embryos. It is thought that the impact of a photoelectron on the macromolecule, forming the gene, might produce, at that point, a structural change that would appear as a hereditary characteristic. A macromolecule possessing 10^n atoms may, theoretically, give rise to $10^n!$ different configurations. Although this figure is considerably reduced by the scarcity of significant atoms, it nonetheless remains extremely large. It is sufficient to recall that there are 3.6×10^6 different ways of grouping 10 persons around a table. It is not enough that the relationship existing between a similar molecular modification and the eye color of a *Drosophila* is totally unintelligible. We shall soon see why.

A macromolecular lesion leading to a mutation and resulting from the appearance of a "hot point," to use Dessauer's expression, may also be caused by thermal molecular agitation, since kinetic theory assigns no limit to rare fluctuations. The mean molecular energy, at absolute temperature T, is given by Boltzmann's equation $(3/2) \times KT$. Schrödinger (10) thus calculates the probable time t that would be required to produce a mutation of threshold energy W. This time is given by the expression:

$$t = \tau e^{W/KT}$$

τ being a constant with a value between 10^{-13} and 10^{-14} sec. At ordinary temperature ($T = 300°K$), for various values of W, expressed in electron volts; we have:

W:	0.9	1.5	1.8 (ev)
W/KT:	30	50	60
t:	0.1 sec	16 months	3.10^4 years

The quantum 1.8 ev corresponds also to a red radiation of wavelength:

$$\lambda = \frac{12{,}350}{1.8} = 6850 \text{ Å}$$

Mutations which would be capable of "reducing species to powder," if they were frequent, remain exceptional, since the intensity of natural ionizing radiations and cosmic rays is very low. Their frequency does not exceed 10^{-4} or 10^{-5} per generation, that is to say, the probability of their founding a family is very low. The stability of a species requires that it remains an exceedingly rare exception for their effects to persist. When a living creature, such as a fish, for example, produces annually 10^7 descendants, all save one are doomed to disappear since the number of individuals making up the species remains statistically constant aside from minor fluctuations. Heredity is the consequence of fixity. The effect of the mutations is blotted out. *Drosophila,* which produces 10 generations a year, has not changed since the Eocene period and is identical to specimens preserved in the Baltic amber for 50 megayears. Many species of living fossils are still in existence today.

Through heredity, we see life perpetuating itself like an unbroken chain. Genetics, through the concepts of soma and germen, shows how false the popular conception of filiation by flesh and blood is.

According to Osborn (1902), and M. Caullery, each species would formerly have passed through a juvenile phase of "plasticity," during which it would have been more sensitive to external influences. In this way, the reptiles diversified mainly during the Secondary period, and then afterward remained almost unchanged; also, the mammals differentiated in the Tertiary period, and have hardly altered subsequently. We know that natural radioactivity has always existed, decreasing regularly during the geological periods, but the intensity of the cosmic radiation could have altered considerably during the galactic revolution of the solar system, which took place within a period of 250 megayears.

Many authors considered that the "life of species" was analogous to that of individuals: that they would disappear, like the latter, through spontaneous degeneration and senility. Simpson (11), in his neo-Darwinism, protested vigorously against this concept and attributed the disappearance to the surrounding environment. The polar migrations, in changing climates, may in fact have destroyed a number of terrestrial species by insurpassable seas.

Through strong mutations, which give rise to monsters that found a line, evolution assumes a teratological character. What is more monstrous than Archeopteryx, a reptilian monster presenting, at one and the same time, the characteristics of reptiles and those of birds? Meanwhile, naturalists are divided on this question: while E. Rabaud and E. Guyenot attribute an evolutionary value to monsters, M. Caullery and L. Cuénot deny them this.

A finalist and naive anthropomorphic conception could, for example, compare the bone structure of a whale to the framework of a boat. If the skeleton of a whale were built on a building slip with the aid of "couples" resting on a "keel," according to the rules of naval architecture, it is clear that the whale would be the result of a design. But its genesis is quite otherwise, for it derives, according to Osborn, from the small African shrew mouse, Tupaïa, and its vertebrate organization must be traced back as far as Amphioxus. Evolution is not a "design" which has developed, but it is the free play of dissymmetric actions creating what is improbable. Evolution has proceeded randomly, by "trial and error." All that was possible was born and all that was viable has lived. We recognize the creative role of chance in mutations. In addition, apart from climatic variations, numerous other causes, apparently very insignificant, have played an important role e.g., the emergence of small rodents that destroyed the eggs of the large reptiles of the Secondary period.

Lucretius and Epicurius (341–270 B.C.) were already aware that none of our bodily organs had been created *for* our use but that it was the *organ that created the use.* Man does not have two eyes *in order to* see, but he sees because he has two eyes. This thought was a profound one, because it was based on the fortuitious occurrence of monsters in which organs, themselves perfect, were associated in a disordered manner. The insect's composite eye has nothing in common with that of cephalopods, and that of the vertebrates also has an independent origin. Our eye has been utilized by a host of species before being transmitted to us. In the same way, the *wing* appeared four times in the animal kingdom in completely different forms. To understand orthogenetic evolution, just as to conceive the origin of life, it is necessary to get rid of all anthropomorphism.

The fact of "oriented" evolution, of orthogenesis, ending, for example, in the building of an optical instrument as perfect as the eye, is a problem that has given rise to innumerable controversies because of its apparently finalist aspect. But if nothing of this is seen in normal physics, that is only because the latter is without the necessary complexity. Cybernetics has thrown some light on this problem, in showing how the increasing complexity of a system engenders ever closer *feedbacks* between its various parts. This feedback is achieved by means of enzymes, hormones, and psychism, which is already apparent in the single-celled organism.

As biological phenomena are always associated with physicochemical phenomena, so psychic phenomena are always associated with biological phenomena. The blind work of mutations is coordinated by the psychism. In the same way, there is no human invention in disordered subconscious work, in madness and in dreaming, in the "Brownian movement" of mental images and ideas. Claude Bernard had shown how the living organism is a self-maintained system with many internal interactions: "a living organism is made for itself, and has its own intrinsic laws." He introduced the principle of an internal causality into physiology. According to Wiener (12) "feedback is the secret of life." It is an organizing principle opposed to entropy, an artificial internal "finality." The organization of the living world is the task of this internal causality.

The living creature is not, as Descartes thought, founded on what is determined and on the simple mechanism. It forms a stable self-regulating system. It has the power of making up its own mind. This internal determination depends on the feedback existing between its various parts. It is possible to assume that mutations are selected, at one and the same time, by the external conditions of the environment and by the choice of the living creatures themselves (13), and to their advantage. This internal direction leads toward a progressive and increasingly more extended conquest of the environment by the living world.

This notion of "choice" is one of the fundamental characteristics of life for we already find its first manifestations in enzymic actions, in which the enzyme *chooses* its substrate by chemical means. P. Wintrebert (14) has therefore proposed a chemical theory of evolution. For the supporters of *parasitism* on the other hand, the bacteria are products of cellular degradation and the viruses are degraded forms of bacteria. Thus, according to Boivin, this *regressive evolution* would be effected in accordance with the regression: bacteria, cycle L, rickettsiae, and viruses. The multiplication of the phages at the expense of the bacteria has, however, nothing in common with parasitism, as Michel observed (15). Parasitism brings with it the enzymic apparatus necessary for its existence. For Lwoff (16), evolution also is a continuous retrogradation; physiological evolution, in its specialization, is a regressive orthogenesis. The specialized animals become obligatory parasites. Morphological differentiation is accompanied by a physiological degradation.

We see psychism reaching its apogee at the end of the evolution of the two great phyla, Protostomia, with the social insects, and Deuterostomia, with the birds and social mammals. The insects, numerous and diversified since the Carboniferous period, ceased their evolution 300 megayears ago. They comprise some 700,000 species and are the most ancient inhabitants of the land masses. There no longer seems to be any doubt that this psychism is a consequence of molecular dissymmetry, but we do not understand the mechanism.

Each time that we wish to pass from the macroscopic to molecular and quantic phenomena, we run up against the unknown. As Leibniz said:

"If, by any process whatever, we were made to perceive the movements of the hypothetical cerebral atoms, we should only see corpuscles in motion and not the corresponding thought, the genesis of which would remain, for us, as obscure as before."

Like entropy, the information peters out on the molecular level.

We owe to E. Dubois (1897) a remarkable law codifying psychism in vertebrates. If the mass of the brain is designated by E and that of the body by P, for each species we have the relation:

$$E = KP^{0.56}$$

Thus, each species is characterized by a linear relation on the logarithmic graph: log E, log P, and the magnitude of K defines its mean relative intelligence. Now, Dubois noticed that this cephalization coefficient K varied by multiples of 2, and from this concluded that intelligence had evolved by sudden, discontinuous leaps under the influence of mutations that had caused an extra division of the neuroblasts during embryonic life. The mass of the adult brain is then doubled. This remarkable law has been extended to birds by L. Lapicque, who gave an explanation of the exponent 0.56 by considering it to result from the combination of a law of number and a law of dimension (numbering of the fibers of the phrenic nerve).

The various species of birds are thus classed on three levels, the cephalization coefficient K doubling on passing from the galinaceae to the palmipedes and to the crow parrots. Birds have constructed "plasters" for themselves to replace a broken foot. The nocturnal migratory birds have the notion of time and guide themselves by the starlit sky. In open sea, on the Isle of Ouessant, the birds circle the lighthouse when they are surprised by the mist and set off again as soon as the sky clears.

In Fig. VI-7, from the work of Dubois and Lapicque, we have shown these linear relations for some fossil and present-day vertebrates. On the abscissa, the logarithms of the masses P are given, expressed in grams (from 10 g to 100 tons) and on the ordinates, the logarithms of the masses E of the brain are presented, expressed in grams (from 0.1 g to 10 kg). Each logarithmic straight line is described by its cephalic index K (from 0.5 to 256). In addition, the coefficient of neuroblast doubling is indicated (from 24 to 33). It is seen that all the vertebrates lie between the limits $E/P = 10^{-6}$ and $E/P = 0.1$.

Present-day species possess a brain of greater volume than that of the fossil species from which they derive. Intelligence can thus be expressed numerically from fishes to man. The anthropoid apes and Pithecanthropus represent the last two levels (ten in the vertebrates) that preceded the appearance of man, and the number of neuroblasts doubled at each stage. The case of the elephant is exceptional and worthy of note. It alone is placed on the same level as Pithecanthropus. The elephant is, in respect to the other mammals, what man is in respect to Pithecanthropus. It is the most intelligent of the animals.

Man, all of whose cells are renewed in 7 years, changes neither in personality nor in psychism. He already possesses, during his embryonic life, his 14×10^9 permanent neurons, which only increase in size with age. These *permanent* nerve cells never multiply and cannot be cultivated. Mental images are preserved throughout the whole of life.

Although the nervous center of invertebrates is, from the anatomical point of view, constructed to another plan than that of vertebrates and cannot be directly compared, Lapique showed, in 1907, that the octopus was on the straight line of the batrachians, and the lobster (total of

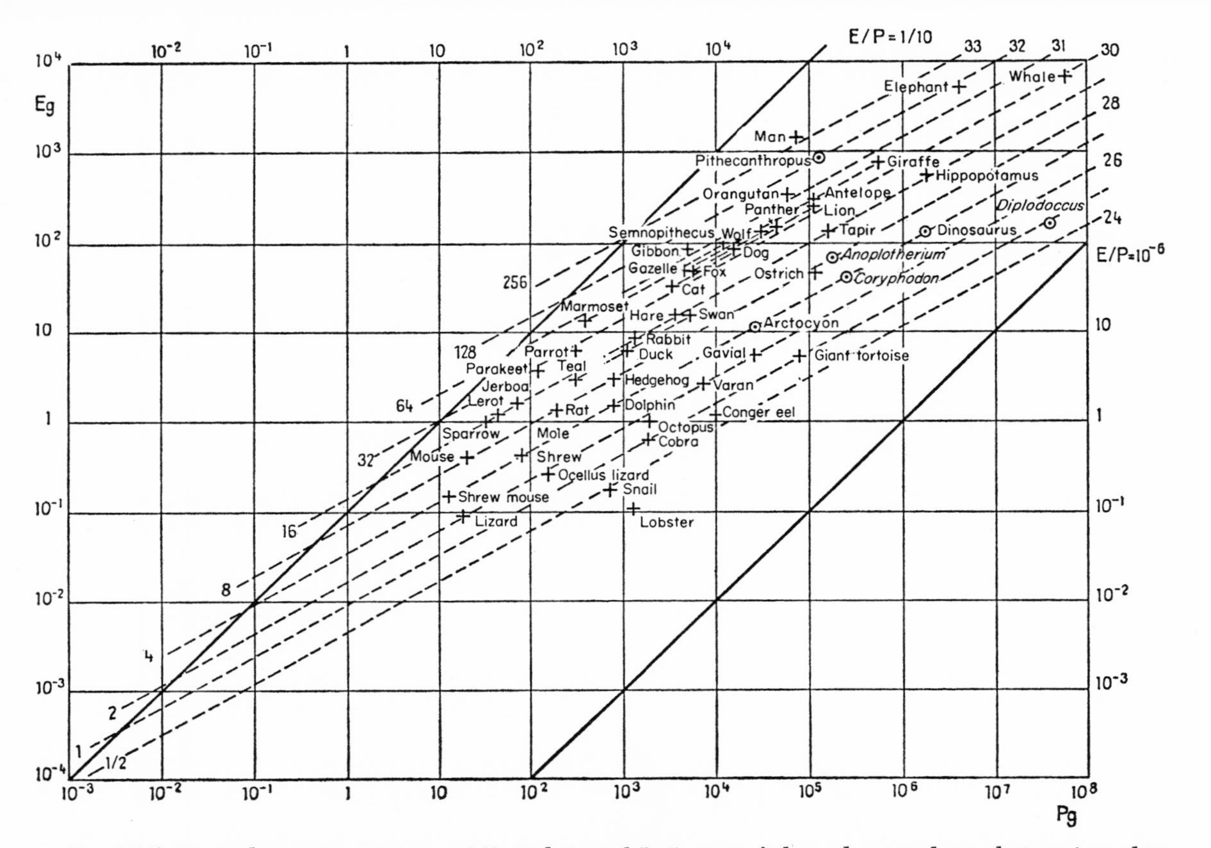

Fig. VI-7. Logarithmic representation of E. Dubois and L. Lapicque's laws showing the evolution of psychism in the living world.

supra- and subesophigeal ganglia are the equivalent of the brain) was at the level subjacent to that of the eel. The cephalopods are, in fact, very highly developed animals, representing the end point of the long evolution of the mollusks, just as the bees and wasps represent the end of arthropod evolution. The Argyronet, a spider that lives in water, has a wonderful, well-known mechanism, the diving bell. Certainly, it works involuntarily (like the majority of human actions), but it was certainly worked out on the first occasion and then *imitated.*

It is noted that, for all vertebrates, the E/P ratio is less than 1/20 and it is hardly possible—for anatomical and physiological reasons—that it would exceed this limit. Now, if one wished to extrapolate Dubois' laws, then for arachnids weighing 1 cg, the ganglia could weigh, at most, no more than a fraction of a milligram, which would place them on the same level as fishes, although their psychism is manifestly superior.

Microcephalic human monsters may occupy a lower brain level, corresponding, for example, to that of reptiles. The study of children brought up by higher mammals (thirty-two known cases, by wolves, bears, leopards, and sheep) has shown the essential role of education between 2 and 5 years. The children became *ineducable* beyond that age. Their psychism was that of Primates without the power of speech, and an inability either to *laugh* or to *cry.* They were unaware of the upright position, and they could only bite, howl, and leap. They fled from daylight but could endure bad meat. They had nothing in common with J. J. Rousseau's "natural man" and Kipling's Mowgli. These facts are of immense philosophical importance. On the other hand, psychism reaches its highest development in the higher vertebrates and social insects, who possess a brain of relatively large volume with many convolutions.

Thus, in the course of evolution, psychism has not developed uniformly in the entire living world, but mainly in the mollusks, with cephalopods, the arthropods, the social insects, and the vertebrates, with birds and mammals. The intelligence of arthropods has been denied. Rabaud (17) cites the case of the spider spinning its silk around the prongs of a vibrating tuning fork, as proof of the automatism of instinct. However, in regard to this, only statistical observation of a large number of spiders, and a great many experiments should provide evidence. If, *in a few cases,* the spider refrains from enveloping the tuning fork, the automatism thesis breaks down.

References

1. H. A. Krebs, *Endeavour* **16**, 125 (1957).
2. H. Devaux, "La nature des particules essentielles de la cellule." Delmas, Bordeaux (1933).
3. A. Dauvillier, *Bull. Acad. Belg. Cl. Sci.* 947 (1963).

4. R. Wurmser, *Bull. Soc. Chim. Biol.* **5**, 305 (1923); *Arch. Phys. Biol.* **1**, 33.
5. E. Rabinowitch, "Photosynthesis and Related Processes" Wiley (Interscience), New York, Vol. I (1945), Vol. II (1951), Vol. III (1956).
6. J. Carles, "Les origines de la vie." Presses Universitaires de France, Paris (1950).
7. H. Munro Fox, *Endeavour* **8**, 43 (1949).
8. F. Holweck, *Compt. Rend.* **188**, 193 (1929); A. Lacassagne, *Ibid.* **188**, 195 (1929); M. Curie, *Ibid.* **188**, 197 (1929).
9. N. W. Timofeeff, *Biol. Rev. Cambridge Phil. Soc.* **9**, 1000 (1934); M. Delbruck, N. W. Timofeeff, and K. G. Zimmer, *Nachr. Biol. Ges. Wiss. Göttingen,* **1**, 189 (1935).
10. E. Schrödinger, "What is Life?" Cambridge Univ. Press, London and New York (1957).
11. G. G. Simpson, "Tempo and Mode in Evolution." Columbia Univ. Press, New York (1953).
12. N. Wiener, "Cybernetics, or Control and Communication in the Animal and the Machine." Hermann, Paris (1949).
13. A. Dauvillier, *Rev. Sci.* **86**, 532 (1948).
14. P. Wintrebert, "The Living Being, Creator of its Evolution." Masson, Paris (1962).
15. G. Michel, "Cahiers d'tudes biologiques." (1957.)
16. A. Lwoff, "L'volution physiologique," Hermann, Paris (1943).
17. E. Rabaud, "Le Hasard et la vie des espèces." Flammarion, Paris (1953).

VII
THE ENERGETICS OF THE BIOSPHERE

Lavoisier compared life to combustion, but the energy transformations that take place within a living creature are not comparable to those that we can produce. Joule, Kelvin, Helmholtz, G. A. Hirn, and J. Joly recognized, in the last century, that the mechanical efficiency of the human being is much greater than Carnot's cycle would indicate for an equivalent heat engine if the heat of combustion of the nutrients is known. A candle burning 10 g of stearic acid an hour in consuming 20 liters of oxygen supplies 25 cal/sec, equivalent to 105 watts. A 70-kg man consuming largely the same amount of oxygen at rest, liberates 27 cal/sec, that is, 23×10^5 cal/day and is capable of doing 0.5 kwh mechanical work per 8-hour day, which is equivalent to 4.3×10^5 cal, that is, a continuous average mechanical power of 20 watts. According to J. Amar's measurements, an 80-kg man, consuming 32 g of oxygen an hour, liberates, at rest, 29×10^5 cal/day, and, at work, 66×10^3 kg, equivalent to 1.5×10^5 cal/day, 34×10^5 cal/day; his efficiency would be 1/3.

We know of no mechanism that converts chemical energy directly into mechanical work. The physical machines that most resemble biological machines are, perhaps, those based on electrocapillarity, but G. Lippmann's electrocapillary motor worked on the same thermodynamic principle. The living creature is not a heat engine obeying Carnot's principle. It produces mechanical work directly and not via heat. The high efficiency of this conversion, 0.21 for man, according to J. Lefèvre, would require, as Guye stressed, a difference of internal temperature of 78°C. The maximum efficiency ρ of a heat engine is, in fact, given by the expression:

$$\rho = \frac{T - t}{T}$$

where T is the absolute temperature of the hot source, and t that of the cold source. The efficiency is integral if the latter is at absolute zero. When t is the temperature of the human body, +37°C or 310°K, the temperature of the hot source for an efficiency of 1/5, will be 388°K or +115°C, which is sufficient to destroy, and sterilize all living matter.

Now living creatures are isothermic and the majority have the same temperature as their environment. This is true for the African cricket, which, with a weight of 2 g, can fly for 12 hours at a speed of 3 m/sec, using up 75 calories per hour per gram, because of its 0.2 g reserve of glycogen and fats. Its efficiency is very high. These measurements are difficult to make, and they are not calculated exactly. As famous a theoretician of hydrodynamics as Navier condemned aviation in advance by calculating that, in flight, the nightingale uses up 1/13 cv. They have often been compared, wrongly, with heat engines. They are capable of effecting other energy transformations of an efficiency unequalled by human industry. Thus, in the glow worm, bioluminescence would be effected with an efficiency of nearly 99%. In the visible spectrum, it emits more than a black body at 1000°C. It is only in rare cases that physics can effect energy conversions with similar efficiency. Thus, the electric cell converts chemical energy almost wholly into electricity. Only one machine converts light into mechanical work and that is the radiometer of Sir. W. Crookes. It still does not function according to a Carnot cycle since the gas is ultrararefied and the molecular energy is kinetic and nonthermal.

This example of the radiometer shows us immediately that the energy transformations are effected at the molecular level in a region in which the principles of thermodynamics are no longer applicable. The high efficiency of the animated engine requires a molecular coordination. Although chemistry formerly depended on chance molecular encounters and resorted unceasingly to heat to make these encounters more probable and to speed up reactions, the new chemistry of the catalysts brings traces of compounds into play which, forming oriented monomolecular layers, coordinate molecular effects and bring about low temperature reactions with improved efficiency. Living creatures behave in this way.

In the same way that the crystal is able to sort out molecules or ions from the environment in order to build its own lattice, so the living organism can choose the molecules that are useful to it. We have seen that the *levo*-tartaric acid molecule was digested by *Aspergillus niger*, although the *dextro* acid was not touched. *Penicillium glaucum*, cultivated on *p*-tartaric (racemic) acid, left the *levo* acid intact. The oyster must extract CO_3^{--} and Ca^{++} from the marine environment, in order to build its mother-of-pearl shell, which is of microcrystals of aragonite inserted in a conchyoline lattice. In 1847, G. Bischof calculated that, to do this, the oyster had to extract the ions from a quantity of sea water of the order of 2.7 to 6.6×10^4 times its weight. Living organisms can sometimes even distinguish and separate isotopes. Thus, Nier showed that C^{12} was concentrated by plants. Dole *et al.* (1) established the existence of a bacterial fractionation of the isotopes of oxygen; the bacteria selectively utilized O^{16}.

Maxwell visualized, in 1871, a fictitious "demon," on the molecular level, that was theoretically capable of getting around the principles of thermodynamics by separating the fast molecules from the slow without doing any work. He assumed that a closed receiving vessel, filled with an isothermal gas, was divided by a septum that was pierced by a hole fine enough to allow only a single molecule to pass at a time. This orifice was covered by a "trap" which the demon could open or close by sliding it without doing work. *Observing* the molecules moving from various points toward the orifice, the demon would allow the fast molecules to pass through in one direction only and the slow molecules to pass in the opposite direction only. Thus he would soon accumulate fast molecules on one side and slow molecules on the other and would have created, with no outlay of energy, a hot source and a cold source capable of feeding a heat engine, contrary to the principles of thermodynamics. In addition to this he would be able to separate the chemical constituents of the air or make a vacuum on one side and an over-pressure on the other. He would never have *created* energy, since he is borrowing it from the environment, but he would have achieved perpetual motion.

For half a century no objections were raised to this fiction. Eddington suggested that Heisenberg's uncertainty principle would prevent the demon making a choice. Brillouin (2) objected that the enclosed space, being isothermic, was a *black body* in which the isotropic radiation would prevent the molecules from being "seen:" Nothing can be discerned optically inside an incandescent furnace. He has shown the equivalence of *information* and *entropy*. The demon would receive *information* and his entropy would increase. Now, the total entropy must remain constant.

Let us recall that the entropy S describes a thermal state as the quotient of the quantity of heat Q and the absolute temperature T. Its variation is expressed by $dS = dQ/T$. It increases when the temperature falls and would be infinite at absolute zero, where molecular movement ceases. In an isolated system, the entropy can only increase. It can remain constant only if it is the seat of reversible conversions.

Boltzmann explained it by the kinetic theory of gases, as a statistical principle, applicable to a reversible system which is evolving toward more and more probable states characterized by an increasing *symmetry*.

In fact, information is a macroscopic effect requiring organs that do not exist on the molecular and quantic scale. Maxwell's fiction would be better represented by Gibb's dissymmetric molecular porous membranes, that are able to select molecules in a way similar to the semipermeable walls of osmosis. The platinum tube of Villard's osmoregulator only allows hydrogen to pass through. Cellophane is permeable to water vapor only and shuts out the gases of the air. Ffeffer's copper ferrocyanide membranes are permeable to water and impermeable to many salts.

It is possible to establish, for electrodialysis, walls, permeable or impermeable to anions and cations, which can be used for water purification. They require an electric current. However, dissymmetric membranes that separate fast molecules from slow molecules are unknown to us. Living organisms alone are capable of building such membranes because of the asymmetry of their molecules.

It would be possible to imagine a living world that was not utilizing solar light but the internal energy of the environment. The kinetic energy of a molecule is, in gas theory, equal to $\frac{3}{2}\,KT$, K being Boltzmann's constant ($K = 1.38 \times 10^{-16}$ erg per degree) and T the absolute temperature.

The thermal energy of a cubic centimeter of air at ordinary temperature, $t = +27°\text{C}$ ($T = 300°\text{K}$) is thus equal to:

$$W = \frac{3}{2}KT \cdot N = 6 \times 10^{-14} \times 2.7 \times 10^{19} = 0.17 \times 10^{7} \text{ ergs or } 0.17 \text{ joules.}$$

In 1 m^3 of water, taken at the surface of the tropical oceans, this thermal energy would be:

$$W = 0.17 \times 10^{3} \times 10^{6} = 170 \times 10^{6} \text{ joules or 47 kwh.}$$

Of course, it would not be possible to extract all the energy from the environment, which would then be cooled to absolute zero, but a cooling limited to about 6°C would already have taken almost a tenth, and the energy source is infinite. There is no analogy between this imaginary process—lack of adequate filter—and the Claude-Boucherot process which, conforming to Carnot's principle, utilizes the temperature difference between the hot surface water of the tropical oceans (+27°C) and cold abyssal water (+2°C at 1000 m).

In "Twenty Thousand Leagues Under The Sea," Verne propelled the *Nautilus* by means of electric cell whose sodium was taken from the marine environment. What human science cannot do, our great ancestor in life, the electric-eel, has been practicing for millions of years. The eel is capable of producing 250-volt electric discharges for 3 milliseconds by separating Na^+ ions in its sodium concentration microcells. The appearance of the electric organs of fishes has been achieved without an apparent progressive evolution, and seems to be the result of a *strong mutation.* It would have been more elegant to supply the submarine with molecular filters that drew thermal energy from the environment either to create a source of heat that supplied a steam engine functioning according to a Carnot cycle, or, better still, to use the thrust of the boiling water directly for propulsion. The new *Nautilus,* which makes use of a nuclear reactor and steam turbine, seems, despite its admirable performances, costly and inelegant in the face of these theoretical molecular solutions. But as Thomson (3) recognized as far back as 1852, "It is impossible to extract, by *inanimate means,* mechanical energy from matter by cooling it to below the temperature of the environment."

If living creatures thus drew their energy from their environment instead of using chlorophyll photosynthesis, they would still live at the expense of solar radiation, and would only cause their environment to cool locally. In molecular Brownian movement, which is continual, the kinetic energy of the agitated granule is also taken from the ambient thermal energy. In conformity with the principle of the conservation of energy, the environment is cooled locally when the granule rises against gravity.

Thus, because of the solar heat, living creatures would be capable of achieving a sort of "perpetual" movement comparable to the Brownian movement. Such autotrophic creatures would have no need of either oxygen, or light, or nourishment. They would take their matter from the ions of the environment, once and for all.

If living creatures do not use this source of energy, they resort still less to nuclear energy. Radioactivity is fatal to them for they are destroyed by any quantum of energy above a few electron volts. There is no place in the energy balance sheet of the living organism for some sort of "vital energy" or for any mysterious radiation. Blondlot's N rays (1905), said to issue from certain regions of the human body, have been attributed to illusions of a psychophysiological nature. Gurwitsch's ultraviolet "mito genetic" rays (1930), which would have accompanied mitosis—studied by Rajewski in 1931 with the photocounter—have not been further confirmed. Nevertheless, mitosis suggests the existence of unknown forces. The pattern of Fig. I-5 suggests a field of force orienting the bacteriophages towards the bacteria, but the orientation may be *post mortem* and due purely to physicochemical effects. The forces invoked by Jordan (4) and Winter's field (5) do not seem, so far, to have been confirmed by observation.

Apart from bioluminescence, which is located in the visible spectrum and is the result of oxidation phenomena that can persist after death, living creatures radiate only as a black body at the same temperature. The human body at 37°C thus radiates a continuous electromagnetic spectrum (3 to 70 μ) whose maximum is situated in the infrared at 10 μ. If its surface area is about 2 m^2, it radiates, according to Stefan's T^4 law:

$$W = 2 \times 10^4 \times 5.67 \times 10^{-5} \times (273 + 37)^4 = 10^{10} \text{ ergs/sec}$$

or

$$10^{10}/4.2 \times 10^7$$

that is, 238 cal/sec. This radiation would be that from a naked man in an enclosed space at absolute zero. The respiratory metabolism would be some 10 times less, and it would soon bring about death. It is compensated for by the radiation of the environment and prevented by fur, fat, feathers, and clothing.

The role of the biosphere, from the point of view of the thermodynamics of the globe, has been discussed a great deal. According to Brunhes (6), "Living organisms have the role of slowing down the degradation of the energy in the world." Chlorophyll synthesis certainly has the effect of slowing down the degradation of energy at the earth's surface. The energy of the photon is converted into the chemical energy of endothermal organic compounds instead of being degraded directly into molecular agitation, that is to say, into heat. But heterotropic creatures, the parasites of the plant kingdom, consume just these energy reserves when producing heat. They reestablish the equilibrium and the degradation of energy. It is, for example, the evaporation of water at the surface of the oceans under the influence of solar heat, and its accumulation in the form of potential energy, that is capable of effecting mechanical work with great efficiency, on mountain summits, in the form of glaciers, and of water retained in high altitude lakes. But in every case, degradation is only slightly *delayed.*

The energy yield of the biosphere is, moreover, extremely low. According to Le Chatelier (7) and Duclaux (8), granting a mean fixation of 100 g of carbon per square meter of vegetation per year, which can release 800×10^3 calories by combustion, the yield would only be 4×10^{-4} of the energy received from the sun (2.1×10^9 calories m^{-2} $year^{-1}$).

According to Rabinovitch, the yield from the ocean plankton would be greater: It would fix 375 g of carbon per square meter annually. Over the whole globe, this fixation would rise to 1.74×10^{17} g of carbon per year, corresponding to 4×10^{17} g of living matter, that is to a fiftieth of the mass of the biosphere.

According to the work of Angot and Milankovitch (9), the mean solar energy received annually per square centimeter of terrestrial surface in our latitudes (40°N) is a tenth of the solar constant (2 cal cm^{-2} mn^{-1}). It is, therefore, (2/10)(1/2)10^6, that is 10^5 cal cm^{-2} $year^{-1}$. It is equivalent to the heat produced by the combustion of $10^5/8.1 \times 10^3$ or 12.3 g/cm^2 of carbon per year. If the chlorophyll function has a yield of 10^{-3}, the amount of fixed carbon would be some 0.012 g cm^{-2} $year^{-1}$. Estimating the mean surface density of the biosphere at 4 g/cm^2, containing 0.4 g/cm^2 of carbon, it is seen that its annual rate of renewal would be 0.012/0.400, that is, 3%. In the laboratory, it is possible to obtain a yield of 10%, 100 times greater.

It is readily seen that the energy requirements of an animal, especially a warm-blooded animal, could not be ensured by chlorophyllic symbiosis, although chlorophyll function is sufficient to ensure those of the plant.

The evaporation of water at the globe's surface absorbs an incomparably greater quantity of energy, estimated at 0.125 cal cm^{-2} mn^{-1}, which is restored to the atmosphere by condensation.

The amount of free carbon of biogenic origin existing in the sedimentary terrains should have been of the same order of magnitude as the amount of free oxygen present in the atmosphere. Since all living matter arose from atmospheric carbon dioxide, 1 g of oxygen corresponds to 0.375 g of carbon. Now, the mass of atmospheric oxygen is $0.21 \times 5.2 \times 10^{21} = 10^{21}$ g, whereas the mass of exploitable fossil carbon in coal, lignites, and mineral oils is estimated to be of the order of 10^{10} g only, instead of 0.37×10^{21} g. Moreover, we have seen that our oxygen was the result of an equilibrium between its production by the chlorophyll function and its consumption by the biosphere, vulcanism, lightning, the combustion of cosmic dust, and so forth, so the amount of fossil carbon should still be greater than 10^{21} g. This free carbon still exists, but only as traces in all sedimentary rocks and in the metamorphic rocks. Besides the volcanic rocks, all the schists, gneiss, and granites have issued from partly biogenic deposits embedded in geosynclinals. According to F. Clarke, sedimentary rocks contain 2% of free carbon and the Algonkian schists contain up to 20%. Now schists are 6 times more abundant than limestones. If they contained only a hundreth of the free carbon, the carbon would represent 6000 times that of the biosphere and a 2-km thickness of rock, containing a thousandth of biogenic carbon, would suffice to account for the free atmospheric oxygen.

The first living creatures—and this was so for thousands of millions of years—were not fossilized, like those of the later geological periods, because they lacked shells and calcareous or silica skeletons. But we are acquainted with graphite gneiss, bitumenous schistes, black dolomites, and ancient graphitoid limestones containing traces of carbon. The partial carbonization of living matter was effected at about 300°C at the bottom of the geosynclinals, out of the reach of air. Carboniferous schists, subjacent to coal seams, contain more carbon than the coal itself, but this finely divided carbon is not exploitable.

According to Gréhant, a good example of the biosphere is obtained in a hermetically sealed aquarium which contains fish living symbiotically with algae and bacteria in an aqueous environment that consists in carbon dioxide, oxygen, and dissolved salts.

In 1900, the Berlin Botanical Gardens possessed a cactus which for 7 years had grown in a sealed glass tube. The earth contained fungus spores that germinated, spread a green layer on the inner walls of the tube, and upon dying gave off carbon dioxide. Thus we find that, in such an isolated environment, according to the principle of conservation of energy, life can be continued only when there are contributions from an outside energy. The closed system is an isothermal transformer of luminous energy into chemical, mechanical, even luminous and electrical, and, finally, thermal energy. The analogy with Crookes' radiometer is complete: This is not a thermal device, but a molecular and quantum one.

Thus, we conclude that efficiency would be slightly more improved if the fish were luminescent, if the aquarium walls were perfect reflectors, and if the light were only admitted through a narrow opening. However, the light is finally reduced to heat. On the other hand, in Pasteur's sealed flask, devoid of air and left in darkness at constant temperature, anaerobic fermentation could not persist indefinitely. The development of yeast was soon arrested and the food supply became exhausted.

References

1. M. Dole, R. C. Hawkings, and H. A. Barker, *J. Am. Chem. Soc.*, **69**, 226 (1947).
2. L. Brillouin, "Science and Information Theory." New York (1956).
3. W. Thomson, *Proc. Roy. Soc. Edinburgh* (1852).
4. P. Jordan, *Physik. Z.* **39**, 711 (1938); *Z. Physik.* **113**, 431 (1939).
5. J. Winter, *Compt. Rend.* **230**, 626 (1950); **232**, 1076 (1951).
6. B. Brunhes, "La dégradation de l'énergie." Flammarion, Paris (1908).
7. H. Le Chatelier, "Leçons sur le carbone." Hermann, Paris (1908).
8. J. Duclaux, "La chimie de la matiere vivante." Alcan, Paris (1910).
9. M. Milankovitch, "Théorie mathématique des phénomènes thermiques produits par la radiation solaire." Gauthier-Villars, Paris (1920).

VIII
THE GEOCHEMICAL ROLE OF THE BIOSPHERE

Four thousand million years ago, before the emergence of life, our planet presented a different aspect from what it does today. The atmosphere was similar to that of Venus, being formed of nitrogen and carbon dioxide, devoid of oxygen, and containing yellowish white clouds of ammonium nitrite. Undoubtedly, the primitive oceans were already as they are today, but the coral reefs, barriers, and atolls did not exist. The continents presented a very different aspect. Of course, this was no longer the lunar volcanic aspect prior to the condensation of the hydrosphere, but it was still rocky and bare. The sedimentary terrains were different from those of today. At the present time, the sediments represent 40×10^6 km^2, that is, a quarter of the continental surface. According to Kuenen, the mean depth of the continental sediments is some 1.5 km, and that of the oceanic sediments three times greater, that is, 3 km of dry mud. They contain 5% biogenic limestone, fossil fuels, and new rocks. Although the mass of the biosphere is less than a hundredth of that of the atmosphere, it takes its matter from the hydrosphere and the atmosphere, and chemical elements pass through it so rapidly that it plays an important geochemical role. It has largely created our present atmosphere and soil, and brought about the formation of numerous rocks. It is responsible for the cycle of the biogenic elements and it effects isotope separations.

All the chemical constituents of living matter pass through similar geochemical cycles. We shall briefly recall the most important; those of carbon, nitrogen, oxygen, phosphorus, sulfur, calcium, manganese, and iron. Some elements, such as boron and copper, are *dispersed* by living matter. Others, such as sodium, iodine, and bromine, are concentrated by marine plants; potassium and rubidium, are concentrated by terrestrial plants; and sulfur, phosphorus, calcium, and silicon are concentrated by marine organisms and bacteria. Iron and manganese are concentrated by bacteria.

Before describing these cycles, it is important first of all to estimate the mass of living matter. This estimation is difficult. Vernadsky (1) believed the figure to be of the same order of magnitude as the mass of free atmospheric oxygen, that is, 10^{21} g. But we have seen that the latter was the result of an equilibrium between its rate of production, by chlorophyll action, and its consumption by the biosphere, vulcanism, and lightning. If this calculation were exact, the mean surface mass of living matter would be of the order of $1/5 \times 10$ m of water, that is 2 m thick, which is clearly much too large. With the seas covering 75% of the earth's surface, it is probably the oceanic *plankton* ("wandering") that represents the major part of the mass of the biosphere. It is mainly located in the surface waters to a depth of some tens of meters. It shows variations with the season and with latitude, and is particularly abundant in the phosphorus-rich cold waters of high latitudes. During the cruise of the *Valdivia* 300 g of living matter per cubic meter of water, containing 8×10^9 Diatoms and 1300 fish eggs, were found at Fisch Bay. For an average 10-m plankton layer, this mass would be some 0.3 g/cm^2 of marine surface. According to Foxton (2), in the Southern hemisphere, from 30° to 70° latitude, the number of zooplankton is maximal between 30° and 55°. Between the surface and a depth of 1000 m, it has a total volume of some 11 cm^3/cm^2 in winter and some 13.7 cm^3 in summer. During the *Galathea* expedition, Steeman-Nielsen (3) measured the rate of production of organic matter in various oceans, using radioactive C^{14} as a tracer. He found an average carbon fixation of 0.123 g per day per square meter, but this figure showed large geographical variations, as it was only 0.020 g at Teneriffe and reached 2 g, that is, one hundred times as much, in the Bengal current. Brouardel and Rinck (4) recently repeated these measurements in the Mediterranean, which is reputedly poor in plankton. The maximum production was observed at about 10 m. Beyond 80 m, where the diurnal luminosity is reduced to a hundredth, fixation becomes negligible. It was 0.030 g of carbon in July and 0.040 g in October per square meter of isolated surface per day (Fig. VIII-1).

Bacterial life, abyssal life, and the benthos, must also be added to these few grams of living matter per square centimeter. Bathyscope descents have shown to F. Bernard rich beds of zooplankton between 200 and 900 m. The marine mud forms charnel houses in which the anaerobic bacteria produce hydrogen sulfide, nitrates, and phosphates. On the continents, the humus of arable land contains plant debris and forms a veritable living environment as a result of the fermentation bacteria and fungi which it contains. It is estimated that a hectare of

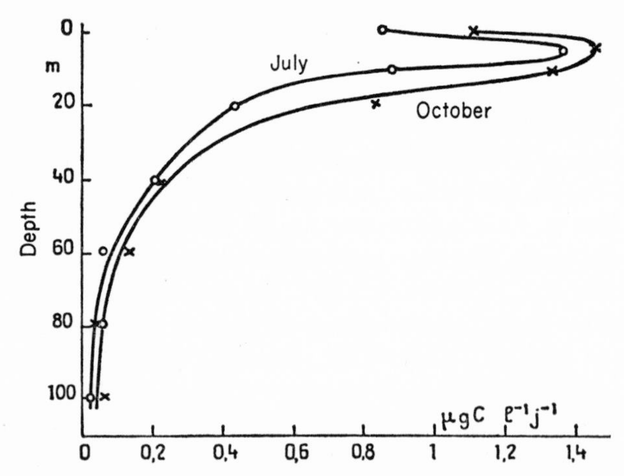

Fig. VIII-1. Production of Mediterranean phytoplankton as a function of depth (after J. Brouardel and E. Rinck) (J^{-1}: per day).

arable land contains ½ ton of these microbes (10^7 per gram of earth). Fixing free oxygen, they convert the plant debris into CO_2 and NH_3, with liberation of heat. Ammonia is, in its turn, oxidized by the nitrifying bacteria into soluble nitrates which, carried by the sap, reform albuminoid substances. Only 5–6% of the globe's surface is almost entirely without life: inland seas, high mountains, and deserts. Nevertheless, the rocks on the highest peaks are often covered with yellow or black coats of lichen. Lichens exist on the final rocks (4746 m) of Mont Blanc and, according to P. Dreux, at 7500 m on Jannu, in the Himalayas. Undoubtedly, Cyanophyceae could have existed at still greater altitudes. Polar snows and ancients firns are sometimes colored by a red-pigmented microscopic alga, *Chlamydomonas nivalis*. The "suntan" of the desert also seems to be of biogenic origin.

Apart from the fresh and marine waters and mud, and the continental surface, the atmosphere is a living environment that has attracted insufficient attention. It contains many particles of inorganic and organic dust and many water droplets, in which germs can live. Pasteur's experiments were the first to detect these microorganisms in the air. At Montsouris, 15,000 bacteria per cubic meter of air and about 130,000 per gram of dust are found. The content rapidly falls, outside inhabited areas, and as a function of altitude. During the ascent of the American manned balloon "Explorer II," only 10 microorganisms, belonging to various species, were collected in 66 m^3 of air at an altitude of 22 km. The pollenic and mycological flora of the atmosphere depend on the meteorological conditions, and above all, on the wind direction.

Flügge and A. Trillat have shown how the organisms could thrive there, because of the food gases and plant and animal *odors*. This biogenic aerosol could be considered as an *atmospheric plankton,* but it is of small mass.

It is not easy to distinguish the living from the nonliving in the muds and terrains, but in estimating the mean surface density of living matter at 4 g/cm^2, that is, 1/250 of the mass of the atmosphere, one should not be far from the truth. The total mass would thus be $4 \times 5 \times 10^{18}$ cm^3 or 2×10^{19} g, that is, $2 \times 10^{19}/1.31 \times 10^{24} = 15 \times 10^{-6}$ of the mass of the oceans. It follows from this that the number of living creatures must be of the order of $10^{19}/10^{-12}$, that is, 10^{31}. Labeled elements, such as radiocarbon, are used to count insects in a particular region, through the intermediate agency of the chlorophyll function. A 70-year-old man has consumed, during his life, 50 tons, that is, almost 1000 times his weight, of food, and 12 million liters of oxygen (17 tons). The substance passing through the life cycle is always the same, so that the amount of substance that has lived must be many times greater than the mass of the oceans and perhaps attain that of the globe. As a matter of fact, if chlorophyll action fixes 1.7×10^{17} g of carbon per year, it produces 4×10^{17} g of new living matter, that is, 1/50 of the mass of the biosphere. If this rate has remained constant for 4×10^9 years, the mass of matter that has lived is 10^9 times that of the biosphere, that is, 2×10^{27} g, or a third of that of the globe, and the maximum number of creatures having lived attains 10^{39}.

Of course, the mass of large living creatures is quite insignificant compared to that of the Bacteria, Cyanophyceae, Algae, and Protozoa. Thus, the mass of the human species, $50 \times 10^3 \times 2.5 \times 10^9 = 10^{15}$ g is only of the order of that of the locust clouds (10^{13} individuals weighing 2×10^{13} g) observed by G. Carruthers in 1889. This estimate was not excessive. The cloud of locusts in Saudi Arabia in 1952 was 15 km long, 8 km wide, and 60 m thick, giving a volume of the order of 10^{10} m^3.

Within their environment, all species are in statistical equilibrium with the climatic conditions and the struggle for existence, which restricts the food supply. But these equilibria are precarious and show variations which are periodic relaxation oscillations and rarer fluctuations. Volterra (5) has given the mathematical laws which govern them:

"When a species is developing in a favourable, limitless environment, the number of individuals of which it is composed is given by an indefinitely increasing exponential function. When the food supply cannot be renewed indefinitely, the coefficient of growth is no longer constant and the population then increases asymptotically. If the food supply is not renewed, or if the medium is polluted by the actual presence of individuals who, because of this, are poisoned, the population reaches a maximum and then decreases to zero."

If the duration of life is equal to the time taken for the population to double when it is increasing exponentially, it is found that the total number of dead is equal to the number of living. If the time constant is T, and the length of life is r, we have:

$$e^{2/T} - 1 = 1$$

As the length of human life (60 years) is close to the time required for the population to double in a number of countries, it can be deduced from this that the number of persons who have lived is of the order of 2.5×10^9.

Rare fluctuations or "explosions" of life are well known—epidemics, floods, etc. Vernadsky quoted the enormous and sudden accumulation of mucilagenous masses found in June 1905 in the Gulf of Trieste. This jelly, floating at the surface of the water, was formed of Peredinians and Diatoms. Falling, in a short while, to the bottom, into an environment devoid of oxygen, these organisms became the prey of bacteria, and produced hydrocarbons.

The Carbon Cycle

The carbon cycle is the most grandiose of the biogeochemical cycles, although this element is not abundant at the surface of the globe. According to Clarke, the terrestrial "crust" contains 4×10^{-3} of carbon, and a fraction of the same order is engaged in the life cycle, making, in total, a few millionths only. We have already retraced the cosmic history of this basic element. All the carbon of living matter derives from atmospheric or dissolved carbon dioxide. Marine algae borrow their CO_2 from bicarbonates:

$$2\ NaHCO_3 \rightarrow Na_2CO_3 + CO_2 + H_2O$$

The neutral carbonate absorbs the dissolved atmospheric carbon dioxide, and reforms bicarbonate. Moreover, the oxygen-nitrogen-carbon dioxide atmosphere extends very deep into the ocean because of the solubility of these gases.

It is *Chlorella,* the unicellular fresh-water alga, that appears to offer the best yield from the point of view of chlorophyllic assimilation. One hectare of *Chlorella* produces 40 tons of dry matter annually, half of which is composed of nitrated substances, and 2.8 tons of fats. One hectare of corn produces only 1.5 tons of dry matter, 180 kg of which are nitrated substances, and 195 kg of fats. *Algoculture,* utilizing water, carbon dioxide, and nitrates, will perhaps be able to ensure the future nutrition of humanity and the synthesis of petroleum, through the fermentation of Algae by the action of anaerobic bacteria, as A. R. Prévôt advocates.

The young carbon dioxide came from active or fossil vulcanism. Biogenic carbon dioxide comes from combustions, fermentations, and the respiration of all living creatures. The soil is a biogenic source of CO_2. According to Reinau's measurements, 1 m^2 of earth liberates 1 g of CO_2 per hour, because of the bacteria and fungi contained in it. As chlorophyll photosynthesis consumes 6.6×10^{17} g of CO_2 per year, the atmospheric carbon dioxide, if it were not rapidly renewed, would be exhausted in $15 \times 10^{17}/6.6 \times 10^{17}$, that is, in 2.2 years only. The ice ages and the luxurious vegetation of the giant cryptogams of the Primary period have often been attributed to a high atmospheric carbon dioxide content. The optimum for life is, as a matter of fact, some 8%, while the air content is only about 3×10^{-4}. In 1878, Schloesing showed the part played by the oceans in regulating the atmospheric CO_2 content. The dissociation pressure of the contained calcium bicarbonate depends only on their surface temperature, which has hardly altered, even during the ice ages. The atmosphere contains $3 \times 10^{-4} \times 5.2 \times 10^{21}$, that is, 15×10^{17} g of CO_2, a mass 10 times less than that of the biosphere. If the carbon dioxide pressure went from 10^{-4} to 5×10^{-4}, the amount of CO_2 dissolved in the ocean would go from 4.5 to 7.3×10^{19} g. The oceans contain 27 times more CO_2 than the atmosphere. It is because of the oceans' stabilizing role that emergent life was not asphyxiated by the volcanic CO_2. During the last century human industry liberated 360×10^9t of CO_2 which, if it had not been absorbed, would have raised the level of atmospheric carbon dioxide by 13%.

Milankovitch has shown that, if carbon dioxide disappeared from the atmosphere, the mean annual temperature of the globe would be lowered by 4°C. If the CO_2 content doubled, the temperature would rise by 5°C. Such variations are out of the question.

In 1844, J. B. Dumas and J. Boussingault were already looking upon living matter as an "appendix" of the atmosphere, and in 1856, J. Moleschott called living creatures: "creatures woven from air by the light."

It was in 1771 that Priestley made the crucial experiment of allowing a candle to burn in an enclosed space until the light was extinguished. Into this space, he then introduced the green parts of a fresh plant and after 10 days exposure to sunlight the air was purified to the point where the candle could again be lit. Lavoisier's famous experiment dates only from 1774. As Moleschott said in relating this experiment: "The sunlit field in which the corn grows, is feeding us. In going through the forest, we feed it with the carbon dioxide which we exhale."

If carbon dioxide is the essential foodstuff of the plant kingdom, a large amount is also fixed by the feldspars of igneous rocks, alumino-alkaline silicates, which make up 60% of the surface rocks, giving rise to the alkaline carbonates and hydrated silica, which forms, with alumina, the kaolin clays:

$$CO_2 + K_2O \cdot Al_2O_3 \cdot nSiO_2 \rightarrow K_2CO_3 + Al_2O_3 \cdot nSiO_2 \qquad +120 \text{ cal/g}$$

But the greatest part of the carbon dioxide is fixed in shallow littoral water, by calcareous microorganisms, Bacteria and Protozoa, in the form of globigerina muds (10% CaO) and, in tropical seas, in the form of corals, whose size is such that they are capable of altering the geography of these regions. According to Deflandre (6), chalk contains up to 10^7 coccoliths per cubic millimeter.

Thousands of species of marine organisms containing calcium exist: mollusks, brachiopods, echinoderms, corals, hydroids, algae, crinoids. Some large formations of oolitic limestones are due to Cyanophyceae and algae.

In 1933, Brussoff showed that the pisolites of hot springs were of bacterial origin. Nesteroff confirmed this and attributed marine oolites to Cyanophyceae.

The calcium carbonate belonging to the land is in soluble form as bicarbonate:

$$CaCO_3 + CO_2 + H_2O \rightleftarrows Ca(CO_3H)_2$$

The latter, a salt of a weak acid and strong base, is hydrolized in the marine environment:

$$Ca(HCO_3)_2 + 2\,H_2O \rightleftarrows 2\,H_2CO_3 + Ca(OH)_2$$

The strong base is finally, dissociated, whereas the weak acid is not:

$$Ca(OH)_2 \rightleftarrows Ca^{++} + 2\,(OH)^-$$

Marine creatures, such as corals, take their calcium from the soluble sulfates. In the absence of life, calcium is not precipitated in the form of carbonate but of sulfate.

Coral shoals represent masses of once living matter, capable of forming mountainous massifs. From the Primary to the Quaternary periods, it is possible to follow the gradual evolution of the coral reefs, from the artic archipeligo of Franz-Joseph land (Silurian) to the Mediterranean basin (Triassic Dolomites) and to the present-day equatorial zone. Suess has shown that the Dolomites should be considered as ancient coral reefs converted into dolomite, $CaCO_3 \cdot MgCO_3$, under the influence of the magnesium chloride of the sea. The magnesium is fixed on the limestone by a protozoan (7), *Trichosoerium sieboldi,* which appears as rods 6 to 10 μ long and 1 to 3 μ wide. Part of the coral reefs returns to the life cycle through the action of perforating Cyanophyceae (*Entophysalis*), secreting citric and malic acids, which form the black above-water zone of the atolls.

Although limestones represent only 5% of the sedimentary deposits, V. M. Goldschmidt estimated their carbon dioxide content at 400 times that of the hydrosphere, biosphere, and atmosphere combined. Only a small part of this carbon dioxide returns to the life cycle. The fossil carbon dioxide, fixed in the biogenic sedimentary limestones, is estimated at $6.56 \times 10^3 \times 5.1 \times 10^{18}$, that is, 33×10^{21} g or 7 times the mass of the atmosphere. It represents only a very small part of the mass of matter which has lived.

It is noteworthy that the atmosphere of Venus also contains at least 450 times more free CO_2 than ours. That appears to indicate that the volcanic carbon dioxide is not retained by the planet and that no biosphere fossilizes CO_2 there in the form of limestones. It is not therefore proper to say, with one of our leading contemporary chemists:

"If a half centimetre more of calcium carbonate had been formed at the surface of the Globe, although the calcareous geological layers are hundreds of metres, there would have been no more carbon dioxide and, therefore, no more solar synthesis. It is on the small quantity of 3×10^{-4} of CO_2 in the air, which has not been absorbed in the past, that present day life on our Globe rests . . ."

Instead, we shall stress, with J. Pompecky (1925), that without vulcanism there would have been no life, and without radioactivity, no internal heat, and, therefore, no vulcanism. On the earth, the two ultimate sources of energy are the solar radiation and radioactivity, which maintain the geochemical activity of the planet, and, hence, life. These two sources are of long duration on the cosmic scale.

Carbon does not escape from the life cycle only in fossil limestones, but also in carbonaceous matter dispersed in rocks, coal, lignites, petroleum, fossil resins, etc. Present-day technical civilization is almost solely based on these fossil combustibles. In dissociating carbonates and in burning these combustibles, human activity returns a considerable quantity of carbon dioxide and carbon to the life cycle.

Outside the carbonates, silica forms the skeleton of a number of marine organisms. The rapid multiplication of Diatoms in the Pacific is due to the abundance of volcanic ashes rich in silica. The valves of Diatoms form the silicic deposits of the infusorial earth of Tripoli and Kieselguhr. In an aqueous environment, silica is also deposited in the form of opal. The travertine deposits of Yellowstone Park are attributed to Cyanophyceae. Jasper is formed by the silica skeletons of Radiolaria. The flint nodules of chalk have a biogenic origin, which is, as yet, unexplained.

Cayeux (8), on finding bacteria in deep sedimentary rocks, stated that their existence in the primitive seas went back to antiquity: "In the light of the facts obtained, one is quite naturally inclined to ask oneself whether the role attributed to inorganic chemistry in the genesis of the sedimentary rocks has not been considerably exaggerated, and whether bacterial action has not exercised a preponderant effect in the formation of non-detritic sediments." The beautiful micropaleontological work of G. Deflandre has shown these conclusions to be valid.

Nitrogen

Although carbon is fossilized in the lithosphere by the biosphere, nitrogen is fossilized in the atmosphere; the natural biogenic nitrates are of little importance geochemically. The mass of atmospheric nitrogen is some 4×10^{21} g. Young nitrogen and the ammoniacal salts are of volcanic origin. This nitrogen is always accompanied by rare gases and argon 40, derived from potassium 40, which establishes its origin at the outset. It is the primitive *cosmic nitrogen* occluded in the magma during its formation. Soufflards of cosmic nitrogen are known, just as soufflards of carbon dioxide and (biogenic) soufflards of methane and hydrocarbons are known. Only one-hundred thousandth of the atmospheric nitrogen is engaged in the life cycle.

Lord Rayleigh (9) has shown the nitrogen of igneous rocks, from the ultrabasic to the acid, to be remarkably constant, 0.04 cm^3/g or 5×10^{-5} g/g of eruptive rock. If this nitrogen was occluded homogeneously in the lithosphere, the latter would contain 50 times more nitrogen than the atmosphere.

Meanwhile, Lepape (10) criticized this conclusion by commenting that this nitrogen contained 25 times more argon than the atmospheric nitrogen. He believed it to be of biogenic origin. This question calls for more research.

An extraordinary thing is that nitrogen, so chemically inert in the laboratory that it cannot even be combined with carbon at the temperature of the arc, that it is absorbed at $\sim$500°C only by calcium, calcium carbide, and lithium, that it enters into combination only when acted upon by high quantum agents such as ultraviolet light, effluvium, and electric discharge, this nitrogen is directly fixed by Cyanophyceae and Winogradsky's bacteria from the root nodules of leguminous plants. The nitrous ferment converts NH_3 into nitrites and the nitric ferment, the nitrites into nitrates.

Animals and some parasitic plants are completely dependent on plants for their protein nutrition. As they must excrete part of this nitrogen in the form of toxic substances, such as urea, the biosphere would very rapidly show a nitrogen deficit without the work of the bacteria which assimilate free nitrogen.

Atmospheric electricity is an important source of ammoniacal and nitric nitrogen. Lightning combines nitrogen and water vapor directly as ammonium nitrite, which is present in rain water:

$$N_2 + 2\,H_2O \rightleftarrows NH_4NO_2$$

The same product is formed by the oxidation of biogenic ammonia by atmospheric ozone:

$$2\,NH_3 + 3\,O_3 = 3\,O_2 + H_2O + NH_4NO_2$$

Lightning also combines oxygen and nitrogen directly, forming nitric oxide, NO, which is oxidized into nitrogen peroxide, NO_2.

Sea water contains 0.2 to 0.5 mg per liter of ammoniacal nitrogen. A liter of rain water in Paris contains 2 mg of ammoniacal nitrogen and ⅔ mg of nitrous nitrogen. Thus, an annual rainfall of 75 cm of water brings 20 kg/ha of combined nitrogen. The air usually contains a proportion of 10^{-8} by volume of NO_2.

In the stormy equatorial regions, this contribution is much greater. In the Congo, 1 to 2 m of annual rainfall fixes 400 kg of nitrogen per hectare. It is estimated that lightning fixes 10^{15} g of nitrogen per year over the whole world. This latter nitrogen returns to the atmosphere when living creatures are decomposed by the action of anaerobic denitrifying soil bacteria that reduce nitrates.

Nitrogen escapes from the life cycle in natural nitrates, in humic substances, in fresh water and sea water mud, and in quinoleins and pyridines contained in coal and mineral oils. Saltpeter, KNO_3, is formed mainly in warm countries at the expense of feldspar rocks. In the Indies and in Egypt, the earth, black and wet during the rainy season, sometimes becomes as white as snow during the dry season and is covered by nitrate efflorescence. The urea and amides of the humus are converted into ammonium salts, into nitrites and nitrates, by Winogradsky's bacteria. In addition, calcium phosphate is produced. Chilean nitrates seem to be due to the oxidation of the guano produced by countless colonies of sea birds, such as cormorants. The nitrates of Tarapaca, which cover immense dry desert stretches, contain sodium nitrate in association with chloride, iodate, and bromate. According to Muntz, they result from the action of sea water on the calcium nitrate produced by nitrifying ferments. The natural nitrates could not be compared to limestones because of their solubility: They make up, at the most, a tenth of the oxygen engaged in the life cycle.

The neutrons of cosmic radiation react with the atmospheric nitrogen. Nitrogen 14, on capturing a slow neutron, gives the radioactive carbon C^{14} which disintegrates to reform nitrogen 14 with emission of β-radiation:

$$N^{14} + n \rightarrow N^{15} \rightarrow C^{14} + p$$
$$C^{14} \rightarrow N^{14} + \beta$$

This radioactive carbon produces active carbon dioxide which, fixed by the chlorophyllic function, *labels* the plants and, consequently, the animals which feed on them.

The half-life of the radio-carbon being some 5568 $\pm$30 years, the carbon of all living creatures contains a very small proportion (10^{-12}) of this isotope in equilibrium. When they die, the isotope, which continues to be converted into N^{14}, decreases in concentration so that a measurement of β activity, carried out on C^{14} enriched by thermal diffusion, enables the date on which the living creature died to be calculated. The method (11), which is applicable for ages between 1000 and 20,000 years, has already permitted numerous identifications in archaeology. It is established that the intensity of the cosmic rays has not changed during this time.

Phosphorus

Phosphorus, which has an abundance of some thousandths in the terrestrial crust, is concentrated to a hundredth in living matter. The source of inorganic phosphorus is apatite, $P_2O_5 \cdot CaO$, proportions of some thousands of which are found in almost all rocks.

This mineral is set free during the decomposition of granites by carbon dioxide. Kuenen showed phosphoric acid to be localized in the cold plankton-rich seas which form the fishing zones. Ammonium phosphate is the source of proteins. Sea water contains 5 mg of phosphates per cubic meter. It loses its population when deprived of nitrogen, phosphorus, and silicon. A seasonal marine biological activity exists. Phosphorus escapes the life cycle by producing extensive calcium phosphate deposits. Those of North Africa, which are regarded as necropolises of tertiary fish, form banks several meters thick, stretching for 4000 km in length, from Tripolitania to Morocco. They contain biogenic uranium which is capable of exploitation. Their existence is explained by an invasion of the fresh waters by the marine environment. After the death of the marine organisms, ammonium phosphate reacted with calcium bicarbonate, precipitating concretions of calcium phosphate. Calcium bicarbonate and phosphate are insoluble in pure water, but soluble in water containing carbon dioxide:

$$(NH_4)_2HPO_4 + Ca(HCO_3)_2 = CaHPO_4 + 2\ NH_4HCO_3$$

The phosphates of the South Sea Islands, such as those of Makatéa, are attributed to the guano of fossil marine birds. Nevertheless, the fact that certain atolls of Tuamotu have their lagoons raised, suggests that they are due to the remains of marine animals which have accumulated by a series of oscillations to which the atoll has been subjected before achieving its final relief. Soluble carbonates, nitrates, phosphates, and sulfates are carried down by the rivers to the oceans and participate in the "salt cycle."

Sulfur

The young sulfur is of volcanic origin. It is contained in the oceans in the form of sulfates. Its biogenic cycle is well known. The plant kingdom extracts sulfur from soluble sulfates to build proteins which are utilized by the whole biosphere. Animal digestion and bacterial degradation produce soluble hydrogen sulfide, which is itself reduced to sulfur by the sulfobacteria, such as *Beggiatoa alba.* This reaction is analogous to chlorophyll action and may be expressed by the symbolic equation:

$$CO_2 + 2\ H_2S = HCOH + H_2O + 2\ S$$

The bacterial oxidation of sulfur by *Thiobacillus* or *Chromatium* produces sulfates again. Moreover, a bacterial reduction and oxidation mechanism, from H_2S to the sulfates and back, exists.

Numerous sulfur deposits, such as those of Texas and Louisiana, thus have a biogenic origin. Naphtha sometimes contains up to 3% of sulfur. Marshy, aerobe-containing soils contain 0.5 to 1% of sulfur. We may cite the famous red lakes of Cyrenaica near El Agheila. The lake of Aïn ez Zania is fed by a source at 32°C. It is covered by a red gelatinous layer, 1 cm thick, formed of sulfur and bacteria (pH: 7.4, salinity: 2.5%, NaCl, $CaSO_4$, H_2S: 20 mg/l). Sulfur is produced by the combined action of reducing and oxidizing bacteria, fixing atmospheric carbon dioxide, and supplying the organic matter of sulfate-reducing bacteria. Two-hundred tons of sulfur are extracted annually. This seems to be the origin of the legends of seas, rivers, and lakes "of blood." (The waters of the Nile changed into "blood"). In August each year, the Alpine lake of Tovel also takes on a blood-red color. But in this instance, it is not a case of sulfobacteria, but of an alga, *Glinodynium sanguineum,* analogous to the well-known *Chlamydomonas nivalis* of the old mountain and polar firns.

The simultaneous production of sulfur and methane by the digestion of drainage mud was suggested. The formation of gypsum from limestone, on building stones, is attributed to oxidizing bacteria.

The two main isotopes of sulfur are $S^{32} = 95.1$ and $S^{34} = 4.2$. Although this ratio is constant in the cosmic rocks (Macnamara and Thode: four siderites and two stony), Thode *et al.* found it to vary in terrestrial sulfur (12). While the Precambrian sulfides and sulfates show the ratio $S^{32}/S^{34} = 22.1$, the present day marine sulfates have 21.8 and the cretaceous sulfides, 23.0. From this the authors concluded that for 800 megayears, a geochemical process has gradually enriched the sulfides and impoverished the sulfates in S^{32} and they identified it as the bacterial oxidation of sulfur and hydrogen sulfide. The oxidizing sulfobacteria would therefore have appeared at this time and would not be primitive, in agreement with our thesis.

Manganese

The geochemical history of manganese is similar to that of calcium, silicon, and iron. It has been very fully described by W. Vernadsky in "La Geochimie." The biogenic importance of this element is due to the ease of transformation of the manganese oxides contained in oxidases, as G. Bertrand has shown. Among the numerous oxides, MnO, Mn_2O_3, Mn_3O_4, MnO_2, MnO_4H_2, and MnO_4H, only MnO, Mn_2O_3, and MnO_2, which is pyrolusite, the most important, are found at the surface of the globe. The young manganese is carried to the surface by vulcanism and orogenic movements. In a humid environment, this mineral is converted into psilomelanes and wads by bacterial action. They are complex hydrated colloids containing iron, calcium and barium. The wads form large manganese concentrations in the ocean depths. Abyssal red clay is rich in manganese. According to J. Murray, rivers carry some 8×10^{12} g of manganese to the oceans annually, but sea water contains only traces of manganese.

Although sea water contains only 50 mg of iron and 5 mg of manganese per cubic meter, red clay contains 6.7% of iron and 0.76% of manganese, that is, a concentration 1.5×10^6 times greater. Correns (13) has given a satisfactory account of the biological extraction of these elements from sea water but Pettersson (14) criticized his conclusions and attributed these concentrations to submarine vulcanism. The element seems to be fixed by marine life in insoluble wads. The Pacific nodules contain 25% manganese, 15% of iron, copper, nickel, and cobalt. The wads are also deposited in the epicontinental seas, lakes, and marshy soils. The black coats of rocks and the "suntan of the desert" are due to such manganese-rich coatings.

Living matter borrows its bivalent manganese from manganese bicarbonate and manganese sulfate solutions. Manganese, like iron, is an active oxygen carrier. Certain fresh water bacteria, such as *Crenothrix* and *Leptothrix,* contain up to 7% manganese. These are autotrophic bacteria, similar to Winogradsky's ferrobacteria, which oxidize bivalent manganese, that is, MnO, into trivalent manganese, into Mn_2O_3 and MnO_2.

Iron

The geochemical history of iron closely resembles that of manganese and its cycle is intimately associated with it. The most important sedimentary iron deposits, those of Lorraine and of Kertsch, are of biogenic origin. According to A. Leclère, the same is true of the iron minerals of Western France. The ferrobacteria oxidize ferrous salts into ferric according to the reaction:

$$2\ FeCO_3 + 3\ H_2O + O = Fe_2(OH)_6 + 2\ CO_2$$

We have seen the first red rocks appear at Algonkian under the influence of life. A great many marcassite (FeS_2) stones are found in chalk, just as flint nodules are found, and their biogenic origin does not seem to be in doubt. However, the genesis of these radiating concretions of iron sulfide is still unexplained.

The minerals of oolitic limonite from Kertsch contain up to 9% manganese. Limonite, which is $2\ Fe_2O_3 \cdot 3H_2O$, results from the weathering of glauconite, the hydrated ferrisilicate of iron and potassium, $KFe^{+++}(SiO_3)_2$. Grains of greenish glauconite, of biogenic origin, are formed at the bottom of the epicontinental seas, at depths below 2 km. They result from the conversion of kaolin alumino-silicates and of the potassium alumino-silicates of micas, feldspars, and chlorites. These greenish colloidal muds, which contain organic matter, have the geochemical interest of extracting the marine potassium in an insoluble form. Thus, sea water contains only 0.5 g of KCl for 30 g of NaCl per liter, while the cosmic abundance of these two alkalis is of the same order. Sea water contains 5×10^{-6} iron and 10^{-8} of nickel. Although the presence of nickel and cobalt is characteristic of the "cosmic iron" of the earth's core and of siderites (cosmic dust), the presence of manganese is characteristic of biogenic sedimentary iron.

It is possible that rubidium and caesium which, like potassium, are unusually rare in sea water, are also fixed in the form of still unrecognized biogenic concretions. So it is that rubidium seems to be concentrated in the oyster's shell.

It has been suggested that limonite is responsible for the reddish color of Mars. It can be seen that no rock could have been more poorly chosen. The reddish volcanic exudates are adequate to account for it.

All these geochemical and biological cycles have taken place, unchanged, for thousands of millions of years. Only for the last few centuries has a new geochemical force appeared with industrial civilization. Fossil combustibles quickly reenter the life cycle: minerals and limestones are reduced; the chemical elements are more and more dispersed. The fossil combustibles, accumulated over hundreds of millions of years, are finally consumed in a few centuries. The geochemical equilibrium of the planet is broken. This is even more true in regard to water flow and in the realm of the living species. Numerous wild species, animal and plant, disappear never to return. Insecticides and artificial antibiotics are disturbing an equilibrium of thousands of years. The destruction of soils, humus, and of many living species will again change the face of the planet, on which virgin nature will have disappeared and where everything will be made for the human species alone, whose proliferation and artificial needs will increase indefinitely. This artificial "conditioning" of the biosphere marks a new phase in evolution, and no example could better demonstrate the influence of life itself on its own evolution, this internal cybernetic feedback—the role of psychism in evolution.

It is not certain that an equilibrium of thousands of years can be broken with impunity in so short a time. A new menace for the living world appears with the large-scale use of nuclear energy. We have seen how life emerged on our planet. It is not certain that it always disappears as the result of a cosmic catastrophe, such as the explosion of a nova, or collision. More probably life disappears as a result of its actual evolution, and by virtue of a fatal destiny: perhaps it is when, after several thousand million years of evolution, an intelligent species—and not necessarily of carnivorous mammals—discovers radioactivity.

References

1. W. Vernadsky, "La Géochimie." Alcan, Paris (1924); "La Biosphère." Alcan, Paris (1929).
2. P. Foxton, *Discovery Repts.* **28**, 191 (1956).
3. E. Steemann-Nielsen, *J. Conseil Perm. Intern. Exploration Mer* **18** (1952).
4. J. Brouardel, and E. Rinck, *Compt. Rend.* **243**, 1797 (1956).
5. V. Volterra, *Compt. Rend.* **203**, 417 (1936).
6. G. Deflandre, "La vie créatrice de roches." Presses Universitaires de France, Paris (1941).
7. M. Gignoux, *Compt. Rend. Soc. Geol. France* 42 (1918).
8. L. Cayeux, *Compt. Rend.* **204**, 1517 (1937).

9. Lord Rayleigh, *Proc. Roy. Soc.* (*London*) **170**, 451 (1939).
10. A. Lepape, *Bull. Soc. Franc. Mineral.* 389–408 (1943).
11. W. F. Libby, "Radiocarbon Dating." Chicago Univ. Press, Chicago, Illinois (1952).
12. H. G. Thode, J. MacNamara, and H. W. Fleming, *Geochim. Cosmochim. Acta* **3**, 235 (1953).
13. W. Correns, *Nachr. Akad. Wiss. Goettingen Math.-Phys. Kl. IIa Math.-Physik. Chem. Abt.* **5**, 119 (1941).
14. H. Pettersson, *Med. Ocean. Inst. Göteborg* 37 pp. (1945).

IX
LIFE IN THE UNIVERSE

The study of the problem of the origin of terrestrial life is the most powerful means available to us of dealing with that of the origin of cosmic life. The problem of the existence of life on a nearby terrestrial planet not only consists of knowing how it has adapted itself to present-day conditions, but above all, of knowing whether it could have existed there in the past. We shall examine first the probability of life in the solar system and then in the galaxy, which is representative of the universe.

We have seen that the emergence of terrestrial life had been dependent on a number of astronomical, geochemical, and geophysical conditions: The mass of the planet had to be sufficient to retain a hydrosphere and an atmosphere, but insufficient to make it a giant planet such as Jupiter, or a still more massive, but hyperdense, microplanet. Its orbit could not have been too eccentric. Clearly, a cometary orbit would be excluded. Its speed of rotation could not be excessive. If the earth were turning 17 times more quickly, gravity would fall to zero at the equator and the hydrosphere and the atmosphere would escape. If the rotation were equal to the period of revolution, our planet would always present the same face to the sun and the oceans and atmosphere would distill and would be frozen on the dark and cold hemisphere. The distance to the sun and the speed of rotation had to be such that the water remained in the liquid state.

If the cosmic radiation had been much more intense, if the sun were a variable star, and if the stellar density in its neighborhood was as high as at the center of a globular cluster, life would have been destroyed as soon as it appeared. In reality, this "world of fireballs rolling towards a threatening destiny" has been very kind to terrestrial life.

If the radioactivity of the terrestrial crust had been very high, the oceans would have remained in a boiling state, and penetrating γ-rays would have prevented life from being born. If seismic activity or vulcanism had been more intense, the same would have applied. Excessive vulcanism would have made the oceans too saline or have covered them with a noxious film of hydrocarbons. If carbon dioxide and oxygen had not been soluble in water and if ozone did not exist, life would never have appeared on our planet. Finally, if the oceans had covered the whole surface of the globe, vertebrate evolution would have been arrested at the fish stage.

All these limitations show us how fragile terrestrial life is and how many conditions, all of which must have operated simultaneously, were required for life to have emerged and to have been maintained. It is not surprising that, of the ten planets forming the solar system, life has appeared on one only. But the Earth is the only known heavenly body covered almost entirely with water, the sole mineral liquid in nature.

The giant planets, whose surface temperature is below —120°C and whose atmosphere consists of ammonia and methane, are manifestly unsuited to life. If the solar radiation increased, or if Jupiter drew as close to the sun as the level of the earth's orbit, the planet would be entirely covered by a fathomless ocean and surrounded by a gigantic absorbent atmosphere of methane, neon, and ammonia. The ultraviolet solar radiation would not even reach the surface of the waters, and because of the absence of asymmetric syntheses no life would be able to emerge there. Of the terrestrial planets, Mercury, as well as our satellite, the moon, are without a hydrosphere and atmosphere and are either torrid or frozen. Only our two neighbors, Venus and Mars, may be considered.

Venus.—The mass of Venus, which is close to that of the earth, is by and large sufficient to have retained the hydrosphere and atmosphere produced by the chemical evolution of its lithosphere. Recent work has shown that Venus' axis of rotation is inclined at the same angle as that of the earth, that is, at 23°, and that its period of rotation is in the region of 1 month. The planet is permanently covered by changing yellowish white clouds that are in suspension in the atmosphere at a level at which the temperature is —40°C. Therefore, nothing is known of its surface. It probably possesses an ocean which is undoubtedly frozen since rotation is very slow. The solar constant there is twice ours but the albedo (0.80) is very high, which means that the radiation reaching the soil is only 0.6 ours. If the Earth received only 60% of the current solar radiation, our oceans, whose mean temperature is no more than 3°C as it is, would be completely frozen. Calculation shows that, if the planet had no atmosphere, the temperature of the sunlight equatorial regions would reach 464°K. The high temperature, 700°K, measured by the "Mariner II" and attributed to the ground of the planet, can, from the principle of the conservation of energy, only be the electronic temperature of its ionosphere.

The nature of Venus' clouds has remained an enigma. Ice cirri are excluded by the yellowish coloration, by B. Lyot's polarization curve, and G. P. Kuiper's infrared spectra, showing the absence of ice and water vapor and the presence of CO_2 alone. Wildt's hypothesis, which assumes that there would be an aerosol of polyoxymethylene hydrates produced from the ultraviolet photochemical reaction of Berthelot and Gaudechon:

$$CO_2 + H_2O \rightleftharpoons HCOH + O_2$$

has not been confirmed by observation of the ultraviolet spectrum. In the gaseous phase and in the presence of more intense solar radiation in the near ultraviolet, photolysis overcomes synthesis. The aldehyde would, moreover, be immediately destroyed by the ozone produced. Finally, oxygen and ozone are not apparent.

The study of the origin of the planetary atmospheres has led us to accept that all the terrestrial planets which have been able to retain an atmosphere, should contain, in addition to carbon dioxide and water vapor, the cosmic nitrogen of Moureu and Lepape—a mixture of nitrogen and rare gases that has issued from their lithosphere. In fact, nitrogen bands have been recognized in the nocturnal luminescence of Venus. According to Kuiper's measurements, the thickness of carbon dioxide above the cloudy layer would be 1 km/atm. On the other hand, Lyot estimated the atmosphere at 1.5 km/atm. This nitrogen would therefore account for 0.5 km/atm. The water vapor is difficult to measure because of the low temperature prevailing at the level of this layer: —40°C according to W. M. Sinton, which would correspond to a vapor saturation pressure of 0.1 millimeter of mercury. According to A. Dollfus, the thickness of water above the clouds does not exceed 0.07 mm. According to J. Strong, the thickness of condensable water is only 0.02 mm.

In these conditions, it is possible to suggest (1) that these clouds are formed of yellowish-white clouds of ammonium nitrite, produced according to Thenard's reaction:

$$N_2 + 2\,H_2O \rightleftarrows NH_4\,NO_2$$

This endothermal reaction occurred under the influence of sparks, of effluvium, and of radiation shorter than 1450 Å, which dissociates N_2. In the presence of an abundant atmosphere of CO_2 absorbing wavelengths shorter than 1800 Å, the aerosol may be produced by lightning. D. Kraus' recent radioelectric observations have shown that Venus is an almost permanent source of "atmospherics" on $\lambda = 11$ m. Since water vapor is scarce at this altitude, this mist, although opaque, is probably almost imponderable. At —40°C, nitrite is stable in the presence of CO_2, but it dissociates at the level of the +50°C isotherm to give nitrogen and water vapor again. Thus, a permanent cycle is achieved under the influence of atmospheric thermic convection.

This fine mist looks like clouds of NH_4Cl produced in gaseous phase. If the NH_4NO_2 crystals are as fine as those of NH_4Cl, they would measure 0.1 μ according to J. J. Trillat. Since they take up water readily, they would be able to form the fine droplets, which Lyot detected, whose index is close to that of water.

The verification of this hypothesis will involve two sorts of observations: The search in the planet's infrared spectrum, for ammonium $(NH_4)^+$ salts ought to show a weak band at 3.2 μ and a strong band at 7 μ. But atmospheric absorption allows 2.6 μ to be only slightly exceeded using the lead sulfide cell. On the other hand, measurement in the laboratory of the polarization and intensity of diffused light as a function of the phase angle, using a flask filled with NH_4NO_2 fumes, should reproduce the curves given by the planet.

Why don't these mists form in our upper atmosphere by photochemical action? It is because of absorption by the oxygen. But, in the beginning, 4.5 thousand million years ago, before the emergence of the biosphere and free oxygen, our planet must have presented the same aspect as Venus; the carbon dioxide was dissolved and combined in the cooled oceans.

If the surface marine waters of Venus are frozen, they do not retain carbon dioxide, and the photochemical syntheses that we have envisaged on the primitive earth do not occur. In this way, the abundance of atmospheric carbon dioxide and the absence of free oxygen is conceivable.

Mars.—The case of Mars is quite different. The planet's mass is 9 times less than that of the earth and 9 times greater than that of the moon. Mars has been unable to retain oceans and able to retain only a very thin atmosphere. The "hydrosphere" of Mars is reduced to a thin cap of hoarfrost which is deposited each year on each winter pole in turn. Its period of rotation and the inclination of its axis of rotation are close to those of the earth, but the solar flux which it receives is only half the terrestrial solar constant. These data enabled a Serbian theoretician, M. Milankovitch, to calculate, in 1920, the theoretical distribution of temperatures at the planet's surface. The temperature of the equatorial regions corresponds to the climate of Franz-Joseph Land, averaging —6°C. But the actual temperature depends, in addition, on atmospheric conditions. Although the atmosphere—reduced to a pressure of 7 mm of mercury (G. Kuiper)—is comparable to that of the earth at about 32 km altitude, the presence of carbon dioxide and the veils of cloud theoretically raise the temperature slightly. Numerous measurements of radiation have revealed a polar night temperature of —60°C. In general, the planet does not reach the temperature of 0°C. Only in perihelion, for a few hours each day, may the temperature momentarily, and in the equatorial regions alone, rise slightly above 0°C. But, since almost all the atmospheric water vapor is condensed in the form of hoarfrost on each polar cap in turn, *neither water nor ice ever exists in these equatorial regions.* In other words, the soil is completely dry. Only polar summer would offer temperatures more favorable to life, but the polar night there is incomparably more severe than on earth.

The water vapor pressure in the atmosphere is very low, but spectrographic research by A. Dollfus detected 0.15 mm of water. Kuiper's work, carried out in the infrared, has confirmed that the polar caps are certainly constituted of frost, which evaporates and sublimates without melting. It can be calculated that Mars' atmosphere is a thousand times drier than ours.

The principal gas that has been recognized in the atmosphere of Mars is carbon dioxide, and it is even more abundant (4.40 m/atm) than on earth (2.20 m/atm), for it is not retained by a hydrosphere. It comes, as on our planet, from soufflards—the last vestiges of an extinct vulcanism. But gases which spectrography cannot detect—such as nitrogen and the rare gases, in particular argon 40, and which, like carbon dioxide, have come from the soil—certainly exist in the planet's atmosphere.

It is possible for the ammonium nitrite clouds found on Venus to form photochemically in the atmosphere of Mars because of the low water vapor pressure. Mars' yellowish clouds and blue haze have, perhaps, been wrongly attributed to soil dust. In fact, about 1 mm NO_2 has been detected.

What does visual and photographic observation of Mars' surface show us? This observation is difficult, since Mars is seen as one would see the moon in binoculars. Maps of Mars, like that drawn up by E. M. Antoniadi, who observed the planet from 1909 to 1940 by means of the large refractor at the Meudon Observatory (the third largest in the world, with a diameter of 83 cm), have only transiently revealed specific details one after the other, depending on the atmospheric disturbance at the time of observation. Large dark expanses have been named "seas," by analogy with those of the moon; the reddish areas are called "continental." Numerous aligned spots are related to the chains of small craters visible on our satellite.

Lyot's polarimetric studies have made a fundamental contribution to knowledge of the nature of the surface of the terrestrial planets. He has shown that the moon, Mercury, and Mars, all possessed a soil of identical nature, which comparative examination has enabled him to liken to volcanic ash. It is known that this ash is very frequently reddish. It is redder on Mars than on the moon because the ferrous silicates were more completely oxidized by the young water vapor at high temperature. Like the moon, the planets passed, during their primitive evolution, through a stage of intensive paleovulcanism, and it is probable that if we could get close enough to Mercury and Mars their surfaces ought to resemble the surface of the moon.

No planet has excited the imagination of fiction writers, and even of a great many ingenuous astronomers, more than Mars. The idea that life was widespread and abundant in the solar system reigned, almost unchallenged, during the whole of the eighteenth and nineteenth centuries. It was partly in order to study Mars that Camille Flammarion founded, in 1882, the Observatory of Juvisy, just as an American amateur, P. Lowell, founded, in 1893, the Flagstaff Observatory on a high plateau (2200 m) in Arizona. Lowell announced in 1884 the presence of a network of "canals" on the planet, probably constructed by intelligent beings, and this unfortunate expression wreaked havoc for 50 years in errant imaginations. It was soon established that Lowell's canals, which were geometrically impossible, were artifacts caused by the instruments used. In 1898, Millochau, Comas Sola, and Antoniadi showed that the appearance of canals was due to chains of small dark spots, that were apparently ranged in line on faults in the planetary surface, as is observed on the moon.

Planetary observations that are made without large instruments should be interpreted with the greatest care. In this respect, the cooperation organized from the second to the twentieth of January 1906 by the French Astronomical Society was instructive. It involved a comparison of independent observations of Jupiter, which were carried out by the most varied methods. Thirty-six observers made up to 17 simultaneous observations, but comparisons of the sketches turned out to be disappointing: Very few details were present on several of them. The sketches differed considerably and most of the details shown did not, in fact, exist.

The Martians, then, had to be abandoned. Meanwhile, in 1876, Camille Flammarion rated the seasonal changes of coloration on the planet, which Liais, Trouvelot, and G. A. Tikhov, attributed to vegetation turning from green to russet red. However, the spectrum of chlorophyll has not been detected by G. P. Kuiper. This presumed "vegetation" was attributed to lichens, which show no seasonal changes of coloration at all.

Like Lowell's canals, Mars' "vegetation" went back to the domain of science-fiction. Kuiper, recently observing the planet by means of the giant telescope at MacDonald Observatory, has shown definitively, with this powerful instrument during the fleeting periods when the atmospheric turbulence permitted perfect visibility, that the dark regions, hitherto assumed greenish, had the *same coloration* as the continental deserts and are merely more dark. Physiological study of the vision of the planet shows that the greenish coloration is due to a contrast effect. Arrhenius had already (2) warned observers against hasty, anthropomorphic interpretations, by showing how the coloration and tonality of the soil depend on the humidity and illumination. It can be added that, if the "seas" are covered by volcanic saline sublimates and saline efflorescences, then their color and tonality may alter reversibly under the influence of ultraviolet light and humidity. We have verified this experimentally in the laboratory with marine salts by reproducing the physical conditions prevalent on the planet.

In the crater of Etna, "fields" of unspotted whiteness and magnificent "prairies" of a beautiful apple green are sometimes observed. These are none other than sublimates of silica, alkaline salts, and ferrous chloride. In Paricutin, deposits of a similar nature are seen which change color with the humidity. As for the C—H bond recognized by Sinton by means of infrared spectrography, we have seen that it is explained by the existence of heterocyclic compounds formed during the paleovolcanic evolution of the lithosphere of the planet.

It can be affirmed that no life exists either on Mars, or on any other celestial body in the solar system, for the simple reason that their atmospheres contain no trace of free oxygen. Any plant transported to Mars would be unable to live there for lack of liquid water, dissolved nitrates, and gaseous oxygen. Furthermore, it would be immediately destroyed by the abiotic ultraviolet solar radiation, to which the atmosphere is transparent, and by the primary cosmic radiation. The lichens, symbiotes of Algae and of Fungi, are in no way primitive, but possess an organization of extreme complexity which required a very long evolution—of the order of thousands of millions of years—in conditions favorable to life.

But if life is impossible on Mars at the present time, does that mean that it did not exist there at a time when the physical conditions were more favorable than today? In other words, could life have been born on Mars, as it was born on earth 4.5 thousand million years ago? Now, the warm marine waters in which life would have been able to appear have never existed on Mars, since the thermal equilibrium which is established at a planet's surface, between the solar radiation that it receives and the infrared radiation of its soil and atmosphere, is immediate. Internal heat of cosmic or radioactive origin plays no climatic role, the thermal conductivity of the solidified magma being very low. So it is that, on volcanoes, snow is seen to persist on a fresh outflow of lava, which is still incandescent just below the surface.

It would certainly be desirable to verify *de visu* these deductions from observation and theory. But it is, perhaps, to be feared that the journey will never take place. If it were carried out at constant velocity, it would require many months and the existence of an astronaut exposed to the cold, to cosmic radiation, X-rays, dogged by thirst, hunger, asphyxia, and insanity. It would, therefore, have to take place at constant acceleration. In 1930, Esnault-Pelterie (3) showed that with a 600,000 cv motor, the journey would then last only 50 hours, which is acceptable, but no organism, it seems, could withstand such prolonged weightlessness. Moreover, no known power would enable the journey to be made. Besides, why must we travel to obtain mineralogical specimens on the planets when meteorites bring large numbers of them to us? The astronautical problem, is not an energy problem, but rather one of the *quantity of movement,* and nuclear energy hardly lends itself to this usage. It is not known whether nuclear disintegration can be directed by a funnel. The easier it is theoretically, with a multistage rocket weighing initially some thousand tons, to leave the earth in a few minutes, the more impossible it is to achieve a prolonged acceleration and to land afterward on planets and return to the earth.

But the astronautical problem can be conceived, more rationally, in the form of a classical, pilotless, teleguided rocket, sending us messages, such as images, by means of a powerful internal source of radioelectric energy, which in that case could be nuclear. The rocket would be able to journey for years with impunity and would not be obliged to return to earth. This engine would be able to go to the planets and send us detailed information.

If, in the solar system, we only find life on earth, that does not mean that it does not exist elsewhere in the universe; on the contrary, the cosmogonic concepts developed in Chapter III that helped us to understand the origin of terrestrial life show that it must be extremely widespread in the galaxy.

We have seen how the formation of multiple stars and planetary systems were the result of stellar interactions occurring in regions of high stellar density, such as the central regions of globular clusters and galaxies. The abundance of couples and red giants has enabled us to calculate the stellar density of the planetary systems, which we have found to be of the order of a hundredth. The galaxy, which comprises some 10^{11} stars, must contain, in a state of statistical equilibrium, some 10^9 planetary systems. But these are not all similar to ours. The characteristic features of our system result from the fact that the two stars in semicollision were largely of the same mass and of the same spectral type. Since interactions can occur between stars of any type and of any mass, the resultant planetary systems may be extremely diverse. For example, they will contain planets with very eccentric orbits, such as cometary orbits, or planets more massive than Jupiter, but reduced to the size of hyperdense asteroids. These planets may not have satellites. Our galaxy would then contain no more than 10^8 systems analogous to ours and there could be as many as 10^7 peopled with higher living creatures.

But these living creatures, whom we shall never know, would differ from us in their isotope composition. They would show an inverse molecular symmetry, like the cosmic creatures imagined by Pasteur. They might—even in our own galaxy—be formed of antimatter, dematerializing into γ-rays in contact with our planet. Because man is theoretically unable to escape from the solar system, the only means that he has of getting to know extraterrestrial life is to exchange radioelectric messages with galaxians of equivalent technical level that are peopling planets belonging to nearby stars.

Such a project has been put into operation by the National Radio Astronomy Observatory (4) under the name of "Project Ozma." They send out and receive directed, rhythmical signals on $\lambda = 21$ cm of the hydrogen atom, the most favorable and probably the most well-known wavelength in the galaxy. For this purpose, a movable paraboloid, 50 m in diameter, has been constructed at Green Bank. Its range would reach 15 light years.

Unfortunately, among the fifty-eight stars surrounding the sun within a radius of 5 parsecs, that is, 16.3 light years, very few lend themselves to such an undertaking. The stellar couples and hyperdense stars are excluded. Among the twenty-six single stars, two only, Tau Ceti (Class G4) and Epsilon Eridani (K2) are of the solar type. They are some 10 light years distant, and the investigations are concerned with them. But the probability of finding living beings there who are familiar with radiotechnics is excessively low, in view of the length of several thousand million years of biological evolution. It will be necessary to wait 22 years for a response to a coded emission. But however low the probability of success, the experiment must be made since it is the only one possible.

Life and thought are perpetual only in a statistical manner and play no role in the dynamic evolution of the universe. Life appears, evolves, and disappears on a myriad of planets following a cycle of some thousand millions of years and is limited in duration by cosmic accidents or by the effect of its own evolution.

REFERENCES

1. A. Dauvillier, *Compt. Rend.* **243,** 1257 (1956); **248,** 1730 (1959); **256,** 836 (1963).
2. S. Arrhenius, "Les atmosphères des planètes." Hermann, Paris (1911).
3. R. Esnault-Pelterie, "L'Astronautique." Lahure, Paris (1930).
4. F. D. Drake, *Sky and Telescope,* **19,** 140 (1960).

BIBLIOGRAPHY

DARWIN, C., "Origin of Species" (1852), London (1950); DARWIN, F., "The Life and Letters of Charles Darwin." Appleton, New York (1898).

MOLESCHOTT, J., "Der Kreislauf der Lebens" (1852).

POUCHET, F. A., "Hétérogénie, ou Traité de la génération spontanée," Baillère, Paris (1859).

BASTIAN, H. C., "The Beginning of Life" (1872), Watts & Co., London (1911); "The Nature and Origin of Living Matter" (1905); "L'origine de la vie." Lamertin, Bruxelles (1913).

ARRHENIUS, S. "Lehrbuch der Kosmischen Physik." Hirzel, Paris (1903); "L'évolution des Mondes." Béranger, Paris (1910).

QUINTON, R. "L'eau de mer, milieu organique." Masson, Paris (1904).

ZSIGMONDY, R., "Zur Erkenntnis der Kolloide," Iena (1905).

COTTON, A., and MOUTON, H., "Les ultra-microscopes et les objets ultra-microscopiques." Masson, Paris (1906).

WALTHER, J., "Geschichte der Erde und des Lebens." Leipzig (1908).

BRUNHES, B., "La dégradation de l'énergie." Flammarion, Paris (1908).

LE CHATELIER, H., "Leçons sur le carbone." Hermann, Paris (1908).

DUCLAUX, J., "La chimie de la matière vivante." Alcan, Paris (1910); "Les colloïdes." Gauthier-Villars, Paris (1920); "Analyse physico-chimique des fonctions vitales." Hermann, Paris (1934).

SCHAFER, SIR E. A., "Life, its Nature, Origin and Maintenance." Longmans, Green, New York (1912).

MOORE, B., "The Origin and Nature of Life." William & Norgate, London (1913); Thornton, London (1919 and 1930).

LUMIÈRE, A., "Le Mythe des Symbiotes." Masson, Paris (1919); "Sénilité et rajeunissement." Hermann, Paris (1932).

PERRIER, E., "La Terre avant l'histoire. Les origines de la vie et de l'Homme." Paris (1920).

OSBORN, H. F., "L'origine et l'évolution de la vie." Masson, Paris (1921).

GUYE, C. E., "L'évolution physico-chimique." Chiron, Paris (1922); "Les frontières de la physique et de la biologie." A. Kundig, Geneva (1936).

COSTANTIN, J., "Origine de la vie sur le Globe." Flammarion, Paris (1923).

VERNADSKY, W., "La Biosphère." Alcan, Paris (1929).

VIALLETON, L., "L'origine des étres vivants. L'illusion transformiste." Plon, Paris (1929).

ESNAULT-PELTERIE, R., "L'Astronautique." Lahure, Paris (1930).

DEVAUX, H., "La nature des particules essentielles de la cellule, micelles ou molécules." Delmas, Bordeaux (1933).

ASTBURY, W. T., "Fundamentals of Fibre Structure." Oxford Univ. Press, London and New York (1933).

HALDANE, J. B. S., "Science and Human Life." Harper, New York (1933).

GUILLIERMOND, A., MANGENOT, G., and PLANTAFOL, C., "Traité de cytologie végétale." Lefrançois, Paris (1933); GUILLIERMOND, A., "Le Chondriome." Hermann, Paris (1934).

MATHIEU, J. P., "La synthèse asymétrique." Hermann, Paris (1935).
BENTNER, R., "Life's Beginning on the Earth." Chapmann & Hall, London (1936).
OPARIN, A. I., "The Origin of Life." Macmillan, New York (1938); Dover, New York (1953).
FINCH, G. I., "La diffraction des électrons et la structure des surfaces." Desoer, Liège (1938).
VALERY-RADOT, PASTEUR, "Œvres complètes de L. Pasteur." Masson, Paris (1939).
KOLTZOFF, N. K., "Les molécules héréditaires." Hermann, Paris (1939).
DEFLANDRE, G., "La vie créatrice de roches." Presses Universitaires de France, Paris (1941).
DAUVILLIER, A., AND DESGUIN, E., "La genèse de la vie, phase de l'évolution géochimique." Hermann, Paris (1942).
SCHRÖDINGER, E., "What is Life?" Macmillan, New York (1947).
GAUTHERET, R. J., "La cellule. Principes de cytologie générale et Végétale." Albin-Michel, Paris (1949).
TÉTRY, A., "Les outils chez les êtres vivants." Gallimard, Paris (1948).
WIENER, N., "Cybernetics, or Control and Communication in the Animal and the Machine." Hermann, Paris (1949).
HAUROWITZ, F., "Chemistry and Biology of Proteins." Academic Press, New York (1950).
BAWDEN, F. C., "Plant Viruses and Virus Diseases." Ronald, New York (1950).
BLUM, H. F., "Time's Arrow and Evolution." Princeton Univ. Press, Princeton, New Jersey (1951).
BERNAL, J. D., "The Physical Basis of Life," Reutledge & Kegen (1951).
DARLINGTON, C. D., "The Facts of Life." Allen, London (1953).
LURIA, S. E., "General Virology." Wiley, New York (1953).
JACOB, F., "Les Bactéries Lysogènes et la notion de provirus." Masson, Paris (1954).
L'HÉRITIER, P., DAUDEL, R., DESTOUCHES, J. L., MONNIER, A. M., LE GRAND, Y., FESSARD, A., MORIN, G., WURMSER, R., AND MONOD, J., "Physique et Biologie." Rev. Opt., Paris (1954).
HALDANE, J. B. S., BERNAL, J. D., PIRIE, N. W., AND PRINGLE, J. W. S., "The Origin of Life." Penguin Books, London (1954).
MORAND, P., "Aux confins de la vie. Perspectives sur la biologie des virus." Masson, Paris (1955).

SUBJECT INDEX

A

Accretion, solar system and, 63–64
Acetaldehyde, synthesis of, 58
Acetylene,
  reactions of, 74
  synthesis of, 21, 70–71
  ultraviolet and, 99
Acquired characters, inheritance of, 135
Adenosine triphosphate,
  Pasteur effect and, 119
  role of, 25–26
Aidiogenesis, 47–48
Alanine,
  formation of, 106
  photosynthesis and, 127
  silk fibroin and, 23, 24
Aldehydes, nitrites and, 105
Algae,
  bicarbonate and, 156
  chlorophylls of, 124
  color of, 126
  evolution of, 133
  geological age of, 117, 118
Alginic acid, nature of, 103
Alkaloids, nature of, 21
Aluminum, occurrence of, 18
Ambystoma, eggs, irradiation of, 135
Amethyst, optical activity of, 112
Amines, pyrochemical reactions of, 75
Amino acid(s),
  formation of, 105
  optical activity of, 108
Ammonia,
  amino acid synthesis and, 106
  cyano compounds and, 73
  formation of, 84
  isocyanic acid and, 74
  living matter and, 17, 19
  ozone and, 95–96
  planets and, 68
  sulfocyanhydric acid and, 75
  ultraviolet and, 96, 99
Ammonium carbamate, urea and, 75
Ammonium chlorhydrate, meteorites and, 49
Ammonium isocyanate, urea and, 75
Ammonium nitrate, formation of, 48
Ammonium nitrite,
  formation of, 92, 96, 105
  Venus and, 170–171
Ammonium oleate, orientation of, 113, 115
Ammonium sulfocyanate, synthesis of, 57
Ammonium thiocyanate, formation of, 75
Amphibians, evolution of, 133
Amphibole, formation of, 73
Amphioxus, evolution and, 137
Anabiosis, freezing and, 40
Angiosperms, evolution of, 133
Animals,
  derivation of, 129
  evolution of, 133, 134
  indispensable elements of, 18
Annelids, evolution of, 133
Anorthite feldspar, formation of, 73
Anthracene, synthesis of, 21
Antimatter, 176
Apatite,
  formation of, 84
  phosphate liberation from, 162
Arachnids,
  evolution of, 133
  psychism of, 142
Aragonite, formation of, 145

Archeopteryx, evolution of, 137
Argon, formation of, 90–91, 160
Arsenic, occurrence of, 18, 19
Arthropods,
  evolution of, 142
  intelligence of, 142
Aspergillus,
  spores, resistance of, 52
Aspergillus niger, tartaric acid utilization by 109, 145
Asphalt, formation of, 79
Asteroids, cosmic dust and, 63, 64
Astronauts, cosmic panspermia and, 52–53
Astronomy, origin of solar system and, 65–66
Asymmetry, living matter and, 20
Atmosphere,
  ammonia in, 96
  escape from, 51
  living matter in, 154–155
  origin of, 88–91, 128
  primitive, 55, 56, 58, 70, 152
  transparency to ultraviolet, 93–98
Autotrophs,
  nature of, 38
  oxygen and, 56
  primitiveness of, 56
α-Azidopropionic acid diethylamide, asymmetric, formation of, 110
Azores, volcanic massif of, 72
Azoxyanisol, polarized light and, 113
p-Azoxycinnamic acid ethyl ether, orientation of, 113, 115

B

Bacillus methanicus, energy of, 56
Bacillus pantotrophus, energy of, 56
Bacillus picnoticus, energy of, 56
Bacillus pyocyaneus, radiations and, 131
Bacteria,
  autotrophic, complexity of, 55
  carbon dioxide fixation and, 158
  chemoautotrophic, primitiveness of, 121
  chlorophyll of, 124
  evolution of, 133
  geochemical energy of, 121
  heredity in, 135
  nitrifying, 90
  reproduction of, 34, 38
  sedimentary rocks and, 160
  size of, 35
Bacteriophage,
  spontaneous generation and, 47–48
  structure of, 32–34
Bacterium hexacarbovorum, energy of, 56
Barium, occurrence of, 18
Basalts, carbides in, 78
Bathybius, nature of, 122
Batrachians, nerve center of, 140
Bee, ectogenesis and, 135
Beggiatoa alba,
  energy and, 16, 56
  sulfur cycle and, 163
Benettitales, evolution of, 133
Benzene, formation of, 71, 74
Bioluminescence, efficiency of, 145
Biosphere,
  chemical composition of, 17
  energetics of, 144–151
  geochemical role of, 152–166
  nature of, 15
Birds,
  cephalization of, 140
  evolution of, 133, 134
Bitumen, formation of, 79
Borides, formation of, 71
Boron, occurrence of, 18, 85
Bromine, occurrence of, 18, 85
Brownian movement, 15
Bryophytes, evolution of, 133
Butyl alcohol
  secondary, dissymetric synthesis and, 112

C

Caesium, sea water and, 165
Calcite, asymmetric syntheses and, 112
Calcium phosphate, deposits of, 162–163
Calcium sulfate, amount in sea water, 106
Camphor, optical activity of, 107
Candle, energy output of, 144
Carbides,
  formation of, 71
  water and, 78–80
Carbon,
  amount fixed by plankton, 153

chemistry of, 20–21
cosmic abundance of, 68
fossil, amount of, 150
isotopic, formation of, 162
retention by earth, 69, 71
Carbonates, meteorites and, 49
Carbon cycle, biosphere and, 156–160
Carbon dioxide,
amount in sea water, 106
fixation of, 81–82, 158
formation of, 76, 78, 80–82, 157
isocyanic acid and, 74
living matter and, 17, 19, 156
Mars and, 172
primitive atmosphere and, 58
respiration and, 118–119
temperature dissociation of, 76
ultraviolet and, 96, 97, 99
Venus and, 170
wet, electric discharge and, 98
Carbon disulfide, sulfocyanhydric acid and, 75
Carbon monoxide,
electric discharge and, 98
formation of, 78, 79
geochemical role of, 79, 80
ultraviolet and, 97, 99
Carbon tetrachloride, formation of, 97
Carborundum, occurrence of, 71
Carotenoids, chlorophyll and, 125–127
Catalysts,
asymmetric, influence on syntheses, 112
energy and, 145
Caytoniales, evolution of, 133
Celestial bodies, capture of, 64
Cell(s), membranes of, 120
Cellobiose, formation of, 102
Cellulose, formation of, 103
Centrosome, radiation and, 131
Cephalopods, eye of, 138
Cephalospid fish, age of, 134
Cerium, occurrence of, 18
Chance, mutations and, 137–138
Children,
reared by animals, psychism of, 142
Chlamydomonas nivalis, habitats of, 154
Chlamydothrix ochracea, energy of, 55–56
Chlorella, yields of, 156
Chlorine,
occurrence of, 18
ultraviolet and, 97
Chlorocruorine, nature of, 123
Chloroform, formation of, 97
Chloromethane, formation of, 97
Chlorophyll(s),
absorption spectra of, 125
appearance of, 55, 122–129
carbon fixed by, 155, 157
energy conservation and, 149–150
intracellular arrangement of, 127
magnesium in, 17
nature of, 122–124
oxygen generation and, 89–90
structure of, 124–125
Cholesteric compounds, orientation of, 113
Cholesterol, meteorites and, 50
Chromatium, sulfur cycle and, 163
Chromite, formation of, 77
Cicutine, synthesis of, 21, 109
Coacervation, living matter and, 120
Cobalt,
occurrence of, 18
optical activity and, 22, 107
Coccoliths, carbon dioxide and, 158
Coelenterates, evolution of, 133
Cohenite, occurrence of, 71
Cold Bokkeveld meteorite, organic material in, 49–50
Colloids, properties of, 22
Comets,
cosmic dust and, 63, 64
spectra, 75
radicals in, 50
Computers, origin of solar system and, 66
Conchyoline, oyster shell and, 146
Contagious bovine pleuropneumonia, virus of, 29
Continents, primitive, 152
Convoluta, symbionts of, 132
Copper, occurrence of, 18
Corals, carbon dioxide fixation by, 158–159
Cordiatales, evolution of, 133
Corn, yield of, 156
Corona discharge tube, organic syntheses and, 98, 105

Corycium enigmaticum, age of, 117
Cosmic dust, solar system and, 63–64
Cosmic panspermia, origin of life and, 49–53
Cosmic radiation,
cosmic panspermia and, 52
evolution and, 136, 137
living matter and, 41–42
Creative chance, origin of life and, 53–56
Crenothrix, manganese in, 165
Creosote, role of, 21
Cricket, efficiency of, 145
Crustacea, evolution of, 133
Crystals, living organisms and, 38
Cuprous oxide, voltaic effect of, 123, 124
Cyanamides,
cyano compounds and, 74
water and, 84
Cyanide, comets and, 75
Cyano compounds,
organic matter and, 57
origin of, 73
Cyanogen,
formation of, 71
nitrogen formation from, 88
Cyanophyceae,
carbon dioxide and, 158
coral and, 159
evolution of, 128, 133
habitats of, 128–129
pigments of, 34
Cycadales, evolution of, 133

D

Daubréelite, occurrence of, 72
Deoxyribonucleic acid,
structure of, 26–28
ultraviolet light and, 41, 42
Desiccation, living matter and, 19
Diamonds, formation of, 79–80
Diatoms,
explosions of, 38
silicon and, 20
Diazomethane, reactions of, 74
Diketopiperazine, formation of, 104
Dinoflagellates,
fossils, meteorites and, 50
Dissymmetry, creation of, 54–55
Distinctiveness, development of, 39
Dolomite, formation of, 158–159
Drosophila, irradiation of, 135, 136
Dwarf stars, origin of solar system and, 66–67

E

Earth,
core of, 70
crust, migrations of, 122
electric field of, 106
first crust, 73, 86
hydrogen and, 69
satellite of, 67
Echinoderms, evolution of, 133
Ectogenesis, 135
Eel, nerve center of, 142
Electric arc, organic syntheses and, 21
Electric discharge, organic matter and, 98
Electric eel, sodium and, 147–148
Electric fields,
living matter and, 42
terrestrial, 106
Electrolysis, organic matter and, 92
Electromagnetism,
cosmic, 62, 63
solar system and, 64
Electrons, living matter and, 42
Elements, cosmic abundance of, 61–62
Elephant, intelligence of, 140, 141
Energetics, biosphere and, 144–151
Energy, life and, 15–16
Enstatite, occurrence of, 73
Entophysalis, coral and, 159
Entropy,
feedback and, 138
Maxwell's demon and, 146
Enzymes, metals and, 18
Epsilon Eridani, 176
Equisetales, evolution of, 133
Erythema, ultraviolet light and, 41, 42
Escherichia coli, radiations and, 131
Ethyl alcohol, synthesis of, 21
Ethyl α-bromopropionate,
asymmetric, formation of, 110
Ethylene,
asymmetric tartaric acid from, 109
ultraviolet and, 97
Etna,
crater, colors of, 174

gases of, 90
Euglena, plants and, 129
Euglena gracilis, structure of, 35, 37
Evolution,
genealogical tree of, 132–134
regressive, 139
Eye, evolution and, 138

F

Fats, synthesis of, 21
Fayalite, formation of, 72
Feedbacks, control and, 138
Feldspar, nitrates and, 161
Fermentation,
anaerobic, development of, 119
energy and, 118–119
Ferrous oxides, as reduction products, 77
Fertilizers, composition of, 18
Filicales, evolution of, 133
Fish, evolution of, 133, 134
Flagella, size of, 35, 37
Fluorine,
occurrence of, 18
retention by sun, 69
Formaldehyde,
electric discharge and, 98
formhydroxamic acid and, 105
origin of, 48
photosynthesis and, 126
synthesis of, 57, 101
ultraviolet and, 97, 99
Formamide, formation of, 103
Formhydroxamic acid, formation of, 105
Formic acid, synthesis of, 21, 57, 74
Fossils,
age of, 117
meteorites and, 50
Freezing, viability and, 40
D-Fructose, ultraviolet and, 99
Fructose diphosphate, formation of, 127–128
Fructose-6-phosphate, Pasteur effect and, 119
Fumaroles, gases of, 90
Fungi,
evolution of, 133
nature of, 134
primitiveness of, 119

G

Gases, living matter and, 17, 18
Genes,
chlorophyll and, 123
nature of, 34
size of, 132
Geothermal energy, organic matter and, 92
Ginkgoales, evolution of, 133
Glauconite,
formation of, 165
marine potassium and, 87
Glinodynium sanguineum, 163
Globular clusters, cosmology and, 65–66
Glucose, structural forms of, 102
Glycine, formation of, 103, 106
Glycocol, polypeptides and, 21
Glycylglycine, formation of, 103
Glyoxal, ultraviolet and, 99
Gneiss, carbon in, 150
Granite,
carbon in, 150
carbon dioxide in, 82
erosion of, 85–86
formation of, 87
fossil-bearing, age of, 117
Graphite,
formation of, 79
protein synthesis and, 107
Gravitation, escape from, 51
Gravitational fields, living matter and, 41
Guaiacol, role of, 21
Guano,
formation of, 161
phosphate in, 163
Gypsum, formation of, 164

H

Halides, primitive, 81
Halogens, living matter and, 18
Heat, living matter and, 39–41
Heat engine, efficiency of, 144–145
Helium, cosmic abundance of, 68
Hematite, age of, 128
Hemocyanin, nature of, 123
Hemoglobin,
nature of, 123
utility of, 129

Heterocycles, epoch of formation, 105
Heterotrophs, nature of, 38
Hornblende, formation of, 73
Hot springs, algae of, 128
Human,
  amount consumed by, 155
  complexity of, 37–38
  energy output, efficiency of, 144
  intelligence of, 140, 142
  number who have lived, 156
  radiation by, 149
  total mass of, 155
Humus, living matter in, 153–154
Hydra, evolution and, 132
Hydrides, formation of, 71
Hydrocarbons, primitive atmosphere and, 58
Hydrogen,
  amino acid synthesis and, 106
  carbon monoxide and, 80
  cosmic abundance of, 68
  electric discharge and, 98
  primitive source of, 71, 83
  retention by sun, 68
  ultraviolet and, 97, 99
  water formation and, 77
Hydrogen chloride, volcanic gases and, 17
Hydrogen cyanide, acetylene and, 74
Hydrogen phosphide, living matter and, 17, 19
Hydrogen silicides,
  stability of, 20
  synthesis of, 70
Hydrogen sulfide, living matter and, 17
Hydrosphere, composition of, 17

I

Ice, planets and, 68
Imidazole, formation of, 75
Industrialization, geochemical equilibrium and, 166
Infusorial earth, origin of, 159
Insects,
  evolution of, 133, 139
  intelligence of, 142
Instincts, olfaction and, 43
Insulin, structure of, 25
Invertebrates, evolution of, 134
Iodine, occurrence of, 18, 85
Iron,
  biosphere and, 165–166
  cosmic abundance of, 68
  occurrence of, 18
  planets and, 69
Iron carbides, carbonaceous meteorites and, 49
Iron oxide, hydrogen and, 77
Isocyanic acid, formation of, 74
Isotopes, selection by organisms, 146

J

Jasper, origin of, 160
Jupiter,
  atmosphere of, 169
  satellites of, 67

K

Kaolin clays, formation of, 87
Kasterite, formation of, 84
Keratin, structure of, 24
Kieselguhr, origin of, 159
Kilauea, gases of, 90
Kinetic theory, cosmology and, 62–63, 65, 68
Krakatoa, blue-green algae and, 128

L

Labradorite, formation of, 73
Lactams, formation of, 104
Lactarium, polyisoprene of, 134
Lactic acid, oxidation of, 119
Lake Kivu, sublimates of, 88
Larderello, gases of, 90–91
Lava,
  soil fertility and, 18
  water in, 81
Leaves, composition of, 127
Lenarto's iron, hydrogen in, 77
Leptothrix, manganese in, 165
Lichens,
  evolution of, 174
  habitats of, 154
Life,
  conditions for emergence of, 168–169
  definition of, 15–16
  emergence of, 121–122
  origin,
    Arrhenius' view, 46, 49–51
    Darwin's view, 46
    Kelvin's view, 46, 49

Pasteur's view, 44
Renan's view, 46
Richter's view, 49
Rostand's view, 46
Teilhard de Chardin's view, 45
Vandel's view, 45
Van Tieghem's view, 46
Vernadsky's view, 46
Virchow's view, 46
von Helmholtz's view, 49
Light, living matter and, 41–42
Lightning,
chemical products of, 48, 92
nitrogen fixation by, 89, 105, 161
Limestone, carbon dioxide in, 159
Limonite,
age of, 128
composition of, 165
Lineola articulata, instincts of, 39
Lipids, photosynthesis and, 127
Lithium, occurrence of, 18
Lithosphere,
cosmic chemistry of, 70–75
nature of, 15
Living creatures,
evolution of, 132–142
nature of, 34–39
Living matter,
carbon of, 156
chemical composition of, 16–20
decomposition of, 17
mass of, 153
rate of synthesis of, 119–120
total mass of, 155
Lobster, nerve center of, 140–142
Lycopodiales, evolution of, 133

M

Macromolecules, living matter and, 22–34
Magma, formation of, 72
Magnesium,
cosmic abundance of, 68
enzymes and, 18
occurrence of, 17
Magnesium sulfate, amount in sea water, 106
Magnetic field,
circular polarized light and, 111
living matter and, 42–43
Magnetite,
cosmic magnetic fields and, 77–78
formation of, 77
Maleic acid, optical inversion and, 109
Mammals, evolution of, 133, 137
Manganese,
biosphere and, 164–165
occurrence of, 18
Marcassite, origin of, 165
Marine environment, primitive, 85–88
Marine mud, life in, 153
Mars,
atmosphere of, 58–59, 83, 88–90, 171–172
canals of, 173
color of, 166, 173, 174
heterocyclic compounds on, 75, 174
lithosphere of, 50, 75
living matter and, 19
possibility of life on, 171–175
surface of, 172–173
temperature of, 171–172
Maxwell's demon, entropy and, 146
Mechanics, origin of solar system and, 66–67
Mediterranean, plankton in, 153, 154
Meioneta, fibroin of, 35
Mellic acid, formation of, 92–93
Mellitic acid, origin of, 48
Membranes, semipermeable, 147
Mercury,
heterocyclic compounds on, 75
lack of atmosphere, 169
lithosphere of, 50, 75
Mercury arcs, energy spectra of, 100
Mesomorphic compounds,
forms of, 113
optical activity of, 115
polarized light and, 113
Metals, occurrence in free state, 72
Meteorites,
composition of, 50, 71–72
fossils in, 50
organic compounds in, 75
origin of, 50
origin of life and, 49
stony, lunar rocks and, 84–85
Methane,
amino acid synthesis and, 106
formation of, 78–80, 82

living matter and, 17, 19
oxidation of, 56
planets and, 68
synthesis of, 57
ultraviolet and, 97, 99
Methanomonas methanica, energy of, 56
Methylene dichloride, formation of, 97
Methylquinoleins, occurrence of, 74
Mineral oil, carbides and, 79
Mitochondria, nature of, 130
Molecular dissymetry, source of, 21–22
Molluscs, evolution of, 133, 142
Monazites, fossil-bearing, age of, 117
Mongolism, cause of, 37
Monsters, evolution and, 137
Moon,
age of, 117–118
heterocyclic compounds on, 75
lithosphere of, 75
optical activity and, 111
silicates of, 72
Mucor,
spores, resistance of, 52
Mugatere,
spores, resistance of, 52
Mutation, evolution and, 135
Myxomycetes, primitive living matter and, 119

N

Naphtha, sulfur in, 163
Naphthalene,
formation of, 71
synthesis of, 21
Nebula, solar system and, 64
Nematic substances, orientation of, 113
Neon, retention by sun, 69
Neutrons, living matter and, 42
New Caledonia, eruptive massif of, 77
Nickel, sea water and, 165
Nightingale, energy use by, 145
Nitrate,
deposits, formation of, 129
natural, nature of, 161–162
Nitrides,
cyano compounds and, 73–74
formation of, 71
occurrence of, 71
water and, 84
Nitrobacter, energy of, 55
Nitrogen,
amount in sea water, 106
biological fixation of, 160
biosphere and, 160–162
cosmic abundance of, 68
fixation of, 89, 90
formation of, 88–90
gaseous, organic matter and, 105
living matter and, 19
oxides, formation of, 48
pentavalent, optical activity and, 22, 107
retention by earth, 69, 71
ultraviolet and, 96
Venus and, 170
Nitrogen dioxide, Mars and, 172
Nitro humulen,
asymmetric, formation of, 110
Nitrosomonas,
complexity of, 121
energy of, 55
Novas, cosmic dust and, 63
Nucleic acids, macromolecular nature of, 25–28
Nylon, monomer, 23

O

Obsidian, earth's crust and, 73
Oceans,
age of, 85–86
carbon dioxide and, 157
formation of, 78, 81
primitive, 152
Octopus, nerve center of, 140
Olber's planet,
lithosphere of, 50, 75
origin of, 67
Oldhamite, occurrence of, 72
Oleic acid, size of, 34
Olfaction, instincts and, 43
Oolites, origin of, 158
Opal, origin of, 159
Optical activity,
origin of, 107–115
Pasteur's views, 108–109
Organic matter,
synthesis, 57–59
pyrogenic, 73–75

Organic substance, meteorites and, 49
Organisms,
  unicellular, evolution of, 129
Organization, living creatures and, 34
Ostracoderms, evolution of, 133
Oxalate, formation of, 48
Oxidation, nature of, 16
Oxides, formation of, 71
Oxygen,
  activation of, 119
  amount in atmosphere, 150
  amount in sea water, 106
  cosmic abundance of, 68
  fixation of, 89
  formaldehyde synthesis and, 101
  formation of, 88–90
  living matter and, 15
  origin of, 16
  photosynthesis and, 126, 127
  primitive atmosphere and, 56–58, 96
  retention by earth, 69
  source of, 72
  ultraviolet light and, 93, 96–97
Oyster,
  shell formation by, 145–146
  rubidium in, 165
Ozone,
  ammonia and, 95–96
  concentration of, 88–89
  cosmic panspermia and, 52
  means of producing, 101
  origin of, 48, 93
  oxidation by, 48
  ultraviolet light and, 93–97

P

Parace, gases of, 87
Paricutin, colors of, 174
Pasteur effect, mechanism of, 119
Penicillium,
  spores, resistance of, 52
Penicillium glaucum, tartaric acid utilization by, 109, 145
Peridot,
  black, formation of, 72–73
  weathering of, 77
Peridotite rock, planets and, 69
pH, living matter and, 18–19
Phenols, role of, 21
Phosphates, amount in sea water, 106
Phosphides,
  occurrence of, 71–72
  water and, 84
2-Phosphoglyceric acid, photosynthesis and, 126–127
3-Phosphoglyceric acid, reduction of, 127–128
Phosphorus,
  biogenic importance of, 16–17, 19
  biosphere and, 162–163
Photographic emulsions, sensitivity of, 123
Photosynthesis, energy and, 16
Phycocyanine, 124
  occurrence of, 127
Phycoerythrin, 124
  function of, 126
Physical agents, action of, 39–43
Physicochemistry, origin of solar system and, 67–69
Pinales, evolution of, 133
Pisolites, origin of, 158
Pithecanthropus, intelligence of, 140, 141
Placoderms, evolution of, 133
Planets,
  composition of, 68
  conditions necessary for life on, 168–169
  envelopes of, 73
  giant, methane of, 82–83
  union of, 67
Plankton, mass of, 152
Plants,
  carbohydrate utilization by, 128
  derivation of, 129
  evolution of, 133, 134
  trace elements in, 18
Plasmagenes, 135
Plasma membranes, function of, 120, 130
Platinum,
  colloidal, properties of, 22
  optical activity and, 22, 107
Platodes, evolution of, 133
Plexiglas, monomer, 23
Polarized light,
  circular,
    dissymmetry and, 110
    production of, 111, 115
  rectilinear, production of, 111

Polyethylene, monomer, 23
Polyhedral disease, nature of, 32
Polypeptides, synthesis of, 21
Polystyrene, monomer, 23
Polytoma uvella, radiations and, 130–131
Polyvinyl chloride, monomer, 23
Potassium,
  isotopic segregation of, 88
  marine, 87
  sea water and, 165
Protein(s),
  macromolecular nature of, 23–25
  photosynthesis and, 127
  synthesis, graphite and, 107
Proteus vulgaris, structure of, 35, 36
Protons, living matter and, 41–42
Protozoa,
  carbon dioxide fixation by, 158
  evolution and, 132, 133
Pseudomonas, hydrocarbons and, 56
Psilomelanes, formation of, 164
Psilophytales, evolution of, 133
Psychism, evolution and, 138–142
Purines,
  formation of, 75
  meteorites and, 50, 75
Pyrazole, synthesis of, 74
Pyridine,
  meteorites and, 75
  synthesis of, 74, 105
Pyrimidines,
  formation of, 75
  meteorites and, 50
Pyrite, occurrence of, 72
Pyroxene, formation of, 73
Pyrrhotite, occurrence of, 72
Pyrrole, synthesis of, 74, 105
Pyruvic acid,
  oxidation of, 119
  photosynthesis and, 127
  reduction of, 127

Q

Quartz,
  circular polarized light and, 111
  inclusions in, 83
  industrial synthesis of, 84
  molecular dissymmetry and, 106–107, 112
  optical activity of, 107, 115

R

Radiations,
  lethal effects of, 130–132
  living matter and, 148–149
  Mars and, 174–175
  mutation and, 135–136
Radioactivity,
  evolution and, 166
  organic matter and, 93
Radiolaria, silicon and, 20
Rain water, nitrogen in, 161
Rare gases,
  loss of, 88
  ultraviolet and, 96
α-Rays, living matter and, 41–42
β-Rays, living matter and, 41–42
Red clay, manganese in, 164
Red giants, formation of, 66
Red lakes, sulfur and, 163
Reproduction, probability of, 31
Reptiles, evolution of, 133, 137
Resins, nature of, 21
Respiration, energy and, 16, 118–119
Respiratory pigments, metals in, 18
Ribonucleic acid,
  composition of, 26
  silica and, 20
Rickettsiae, evolution and, 139
Rickettsia exanthematotyphi, size of, 31
Rivers,
  manganese and, 164
  salts of, 84
Robots, 15, 43
Rock(s),
  carbon dioxide and, 158
  igneous, nitrogen of, 160
  red, emergence of, 128, 165
  sedimentary,
    bacteria and, 160
    carbon in, 150
Rockets, Mars and, 175
Rock salt, primitive halides and, 81
Rubidium,
  occurrence of, 18
  sea water and, 165

S

Saccharomyces ellipsoideus, radiations and, 130–131

Salts,
ultraviolet and, 101
volcanic cycle of, 85–88
Saltpeter, formation of, 161
Santorin, exhalations of, 87, 88
Saturn,
ring, origin of, 67
satellites of, 67
Schists, carbon in, 150, 151
Schreibersite, occurrence of, 72
Scoria, formation of, 70
Sea water,
biosphere and, 17
manganese in, 164
nitrogen in, 161
phosphates in, 162
potassium in, 165
Seismic activity, energy of, 122
Selachians, evolution of, 133
Serpentine, formation of, 77
Siderites, hydrogen in, 77
Silica,
biosphere and, 159–160
formation of, 71, 83
Silicates, formation of, 72
Silicides,
formation of, 71
water and, 83
Siliciformic acid, stability of, 20
Silicioxalic acid, stability of, 20
Silicon,
cosmic abundance of, 68
living matter and, 19–20
minerals and, 15
occurrence of, 18
Silicon sulfide, water and, 83
Silk fibroin, structure of, 23–24, 104
Silts,
clayey, molecular dissymmetry and, 106–107
Smectite forms, orientation of, 113
Sodium,
comets and, 75
electric organs and, 147–148
occurrence of, 18
source of, 86
Solar radiation pressure, cosmic panspermia and, 51–52
Solar system,
origin, 63–65
astronomical problem and, 65–66
mechanical problem and, 66–67
physicochemical problem and, 67–69
Spar,
circular polarized light and, 111
origin of, 112
Spirophyllum ferrugineum, energy of, 55
Sponges, evolution of, 133
Spontaneous generation, origin of life and, 47–49
Stars, interactions, 66
Stearic acid,
combustion of, 144
films of, 120
Stone of Orgueil, composition of, 49, 50
Stromboli, gases of, 90
Strontium, occurrence of, 18
Styrolene, formation of, 71
Sucrose, photosynthesis and, 127
Sugar,
oxidation of, 15
synthesis, 101–102
electric discharge and, 98
ultraviolet and, 99
Sulfides,
meteorites and, 49
occurrence of, 71–72
water and, 84
Sulfocyanates,
formation of, 48
occurrence of, 74
Sulfocyanhydric acid, formation of, 75
Sulfur,
biosphere and, 163–164
energy and, 16
Sun,
composition of, 68
radiation of, 94–95, 106
"Suntan,"
desert, origin of, 154, 164
Supernovas, cosmic dust and, 63

T

Tannins, role of, 21
Tartaric acid,
asymmetric, synthesis of, 109
attempts to make asymmetric, 110
isomers, utilization of, 109, 145
synthesis of, 21
Tau Ceti, 176

Teflon, monomer, 23
Temperature,
carbon dioxide and, 157
chlorophyll function and, 128
Teratology, radiations and, 135–136
Thermal energy,
mutations and, 136
use by living creatures, 147, 148
Thiobacillus, sulfur cycle and, 163
Thiophene, synthesis of, 74
Thiourea, formation of, 75
Thyroxine, intelligence and, 39
Tides, causation of, 64
Titanium, high temperature chemistry of, 71
Tobacco mosaic virus,
parts, infectivity of, 32
size of, 34–35
Tobacco necrosis virus, crystals of, 29, 31
Tomato bushy stunt virus, crystals of, 29
Travertine, origin of, 159–160
Triarylmethyl radical,
chlorinated, asymmetric, 111
Trichosoerium sieboldi, dolomite formation and, 159
Trinity, asphalt of, 79
Triosephosphates, photosynthesis and, 127
Troilite, occurrence of, 72
Tupaïa, evolution and, 137
Turnip yellow mosaic virus, crystals of, 28, 29
Turpentine, optical activity of, 107
Twins,
monozygotic, determination of, 39

U

Ultrasound, living matter and, 41, 42
Ultraviolet light,
chemical reactions and, 97
cosmic panspermia and, 52
creative chance and, 54
deoxyribonucleic acid and, 41, 42
organic matter and, 93, 98, 99, 107
organic syntheses and, 57
Universe,
age of, 62
evolution of, 61–63
life in, 175–176
Uranus, satellites of, 67
Urea,
formation of, 74–75
nitrogen cycle and, 161
synthesis of, 21

V

Vanadium, occurrence of, 18
Van't Hoff's law, 40
Vegetation, Mars and, 173–174
Venus,
atmosphere of, 58–59, 83, 88–90, 92, 159, 170–171
clouds, nature of, 170
living matter and, 19
possibility of life on, 169–171
temperature of, 169–170
Vertebrates,
evolution of, 134
eye of, 138
Viruses,
evolution and, 133, 134, 139
filtrability of, 28–29
replication of, 32
Volcanic gases,
nature of, 17, 72
origin of life and, 50
oxygen and, 56, 57
sulfocyanates and, 48
Volcanos,
exudate, salinity of, 86
manganese and, 164
marine salts and, 87
nitrogen and, 90, 160
Volvox, organization of, 132
Vulcano, gases of, 90

W

Water,
evaporation, energy and, 149, 150
gases dissolved in, 106
hypercritical, salt transport by, 86–88
living matter and, 16–19
Mars and, 171, 172
photolysis, energy of, 123
radiations and, 131
temperature dissociation of, 76
ultraviolet and, 97, 101
Water vapor,
amino acid synthesis and, 106
carbon monoxide and, 80

formation of, 75–78
oxygen formation from, 88
Venus and, 170
Whale, evolution of, 137
Wing, evolution and, 138
Worms, evolution and, 132

X

Xanthophyll, leaf and, 127

Z

Zinc, occurrence of, 18